Process Innovation

Reengineering Work through
Information Technology

Process Innovation

Reengineering Work through Information Technology

Thomas H. Davenport

Ernst & Young
Center for Information Technology and Strategy

Harvard Business School Press
Boston, Massachusetts

Printed in the United States of America

94 95 96 97 10 9 8

The paper used in this publication meets the requirements of the American National Standard for Permanence of Paper for Printed Library Materials Z39.49-1984.

Text design by Wilson Graphics & Design (Kenneth J. Wilson)

Library of Congress Cataloging-in-Publication Data

Davenport, Thomas H., 1954–
    Process innovation : reengineering work through information technology / Thomas H. Davenport.
        p.   cm.
    Includes bibliographical references and index.
    ISBN 0-87584-366-2 (acid-free paper)
    1. Information technology.   2. Technological innovations.
    3. Organizational change.   4. Production engineering.   I. Title.
    HC79.I55D37   1993
    338'.064—dc20   92-21959
                                                                                    CIP

To my parents, with gratitude for their childrearing processes

Contents

Preface

Among the technology-oriented consultants and business academics with whom I associated in the mid- to late 1980s, there was a general sense of great opportunity in the air. The opportunity was to apply information technology to the redesign of business processes. None of us knew exactly what this meant, but occasionally we would find a pioneering company that seemed to have tried it.

Jim Short (then of MIT, now at London Business School) and I decided to capture the efforts of these companies, and some general principles about what worked and what didn't, in an article. It took us a couple of years to find enough examples of process redesign (as we called it then) to make a good article. By the time it was published by *Sloan Management Review* in June 1990, interest in the topic had increased considerably.

When the *Sloan* article was published I was working at Ernst & Young's Center for Information Technology and Strategy, which proved to be a great platform for learning more about this topic. Many firms visited the center for briefings and informal discussions about process innovation (as we began to call it), and I could also learn about the implementation of these ideas by getting involved in various E&Y consulting engagements. We started a multiclient research program on process innovation, in which we could try out ideas and frameworks on a broad variety of companies that were interested or active in the field. The center played an important role in educating E&Y clients about process innovation, and in developing the firm's consulting methodology in this area. Alan Stanford, who did more than anyone else to establish the center (and who hired me to work in it), and Bud Mathaisel, who ran the center and ran interference for its researchers, deserve much credit for making the book and these learning activities possible.

Several members of the center assisted by writing first drafts of chapters: Mary Silva Doctor with a chapter on approach that was later blended into the first section of the book; Alex Nedzel

with the chapter on IT-based implementation of process innovation; Suzanne Pitney with the chapter on organizational change
management; and Larry Prusak with the chapter on information
as an enabler of process innovation.

Other members of the center who researched or reviewed aspects of the book include Jennifer Burgin, Scott Flaig, Chris Gopal,
Jim McGee, Vaughan Merlyn, Phil Pyburn, and Greg Schmergel.
Carol Oulton, Janet Santry, Jean Smith, and Susan Sutherland
were very helpful with graphics or word processing.

Working at Ernst & Young, which had a strong consulting
practice in total quality management, also helped me understand
how process innovation differed from process improvement. I now
feel that these two concepts are poles on a continuum of approaches to operational performance improvement. Jim Harrington, Pravesh Mehra, and Terry Ozan of E&Y were particularly
helpful in this regard.

Several friends and former colleagues at the Harvard Business School read and influenced the book. They include Lynda
Applegate, Bob Eccles, Nitin Nohria, and John Sviokla. Jim Cash
and Warren McFarlan gave me feedback on the original article
that was useful in writing the book.

Carol Franco, my editor at the Harvard Business School Press,
made the publishing process seem easy, made the book seem better, and made all my questions seem reasonable. John Simon and
Natalie Greenberg made my prose much more compact and read-
able. Two reviewers on behalf of the press (Judy Campbell at
Xerox and an anonymous academic) also made valuable suggestions on issues of both content and structure.

A book is a family's project, not an individual's. My wife Jodi
remained cheerful and encouraging through the late nights, early
mornings, and weekends devoted to the book rather than the
family. Even my sons, Hayes and Chase, only occasionally re-
sented not being able to play with me or with the family's best
computer.

I am also extremely grateful to the managers of companies
that were undertaking process innovation initiatives, who shared
their experiences and insights with me quite freely. Process innovation (or reengineering, redesign, and so forth) was invented
not by consultants or academics, but by these bold and intelligent
businesspeople. I simply jumped on their bandwagon at a relatively early stage.

Chapter 1

The Nature of Process Innovation

In the face of intense competition and other business pressures on large organizations in the 1990s, quality initiatives and continuous, incremental process improvement, though still essential, will no longer be sufficient. Objectives of 5% or 10% improvement in all business processes each year must give way to efforts to achieve 50%, 100%, or even higher improvement levels in a few key processes. Today firms must seek not fractional, but multiplicative levels of improvement—10X rather than 10%. Such radical levels of change require powerful new tools that will facilitate the fundamental redesign of work.

The needed revolutionary approach to business performance improvement must encompass both how a business is viewed and structured, and how it is improved. Business must be viewed not in terms of functions, divisions, or products, but of key processes. Achievement of order-of-magnitude levels of improvement in these processes means redesigning them from beginning to end, employing whatever innovative technologies and organizational resources are available.

The approach we are calling for, *process innovation,* combines the adoption of a process view of the business with the application of innovation to key processes. What is new and distinctive about this combination is its enormous potential for helping any organization achieve major reductions in process cost or time, or major improvements in quality, flexibility, service levels, or other business objectives.

Executives in the organizations we studied and consulted for have expressed great interest in process innovation. They have spent much money and time on less structured and less ambitious approaches to business change with little result. The successes of pioneering firms with process innovation initiatives offer them new hope.

1

The early results of process innovation are undeniably attention-grabbing. IBM Credit reduced the time to prepare a quote for buying or leasing a computer from seven days to one, while increasing the number of quotes prepared tenfold. Moreover, more than half its quotes are now issued by computer. Federal Mogul, a billion-dollar auto parts manufacturer, reduced the time to develop a new part prototype from twenty weeks to twenty days, thereby tripling the likelihood of customer acceptance. Mutual Benefit Life, a large insurance company seeking to offset a declining real estate portfolio, halved the costs associated with its policy underwriting and issuance process. Even the U.S. Internal Revenue Service achieved successful process innovation, collecting 33% more dollars from delinquent taxpayers with half its former staff and a third fewer branch offices.

Radical process change initiatives have been called various names—e.g., business process redesign and business reengineering. For several reasons, we prefer the term business process innovation. Reengineering is only part of what is necessary in the radical change of processes; it refers specifically to the design of the new process. The term process innovation encompasses the envisioning of new work strategies, the actual process design activity, and the implementation of the change in all its complex technological, human, and organizational dimensions.

BUSINESS DRIVERS OF PROCESS INNOVATION

That Japanese firms discovered (or at least implemented) process management long before the West (see Appendix B) helps explain their worldwide economic success. Process improvement and management have characterized some Japanese corporate cultures for decades, and enabled firms in a number of industries to develop fast, efficient processes in such key areas as product development,[1] logistics, and sales and marketing.[2] Often, these highly refined processes are introduced with little attendant use of advanced technology or radical approaches to human resource management.[3] They simply are logical, balanced, and streamlined. The Japanese firms that have developed such processes constitute a major competitive driver for the adoption of process innovation by their Western counterparts.

Today, with competition extended to the execution of strategy, firms frequently woo customers on the basis of process innovation.

The objective of the innovation might be process time reduction—in home mortgage financing, for example, timely approval of loans, to the extent that it reduces the period of uncertainty for home buyers and sellers, constitutes a competitive advantage. Citicorp has reduced process time for some mortgage approvals to fifteen minutes (although recent losses in its retail bank suggest that it might also optimize its processes for consistency and risk avoidance).

Process innovation can also support low-cost producer strategies. Companies that eliminate, for example, costly aspects of their product delivery processes can pass the savings on to customers. The life insurance firm, USAA, has a process objective of avoiding the costs of an agency relationship by relying on telephone-based marketing. USAA's telephone customers are consequently paying lower premiums than those offered by other insurers for the same coverage. The Charles Schwab discount brokerage house also uses this approach.

Competitive pressure is not the only driver of process innovation; increasingly, customers are the impetus for radical process change. The automobile and retail industries offer two notable examples of customer-driven change. Auto manufacturers, responding to intense foreign competition in the 1980s, forced suppliers to increase the quality, speed, and timeliness of their manufacturing and delivery processes.[4] In the retail industry, Wal-Mart has established practices of continuous replenishment, supplier shelf management, and simplified communications that have significantly influenced its suppliers, including such giants as General Electric and Procter & Gamble.[5]

Finance is another powerful driver of process innovation. Companies that have assumed heavy debt loads as a result of leveraged buyouts or fending off corporate raiders often need to cut expenses substantially to improve profitability. Process innovation can be more effective at removing unnecessary costs than many other alternatives such as business unit sales and early retirement programs (which result in a company's most employable people leaving to find other jobs). It may be that in the 1990s overburdened companies will come to terms with their debt through operational, process-based restructurings, rather than financial maneuvering.

There are other occasions that provide "fresh start" opportunities for process innovation. For example, it may be desirable to redesign a process that is to be outsourced before turning it over to an outside firm. A merger can provide an opportunity for replacing redundant processes with a newly designed process that better accommodates the new firm's objectives. Even a poor IT infrastructure can be an opportunity for process innovation; many firms today need to rebuild major systems, but they should not construct them to support inadequate or inferior processes.

Process innovation can also respond to the need for better coordination and management of functional interdependencies.[6] Better coordination of manufacturing with marketing and sales, it is reasoned, will allow a company to make only what its customers will buy. In the consumer foods business, this process objective often takes the form of reducing the likelihood that goods will become stale; passing information from sales to manufacturing has enabled Frito-Lay[7] and Pepperidge Farm to realize substantial reductions in stales. The automobile industry has labeled this type of coordination "lean production," an approach cited by a major MIT study as key to Japanese success.[8] Achieving a high degree of interdependence virtually demands both the adoption of a process view of the organization to facilitate the implementation of cross-functional solutions, and the willingness to search for process innovations. Existing approaches to meeting customer needs are so functionally based that incremental change will never yield the requisite interdependence.

Of the many operational reasons that private-sector organizations embark on process innovation initiatives, almost all can be traced to the need for improving financial performance. Process cost reduction translates directly into that objective. Other process objectives, such as time reduction and improved quality and customer service, are assumed to translate into higher sales or less expensive production. Even process objectives that involve worker learning and empowerment are ultimately oriented toward improving financial performance, the assumption being that fulfilled workers will be more productive workers. Because process innovation initiatives consume resources that might be spent in some other way, it is reasonable to expect that they will yield financial benefits. However, as will be shown later, improved financial performance is often problematic as the only stated vision or objective for process innovation. Nonfinancial objectives are normally more likely to inspire vigorous efforts to improve work.

A final rationale for process innovation is that it suits our business culture. Even if Western firms could solve their financial problems and satisfy their customers through incremental improvement, why should they suppress their appetite for innovation? Perhaps what American companies need is a process approach that marries radical innovation and the discipline of continuous improvement. A Western version of quality might focus on results as well as process. U.S. firms need a process management approach that embraces both human enablers of process change and the tool that has changed business most over the past three decades—information technology. Western firms can only hope that the combination of process thinking and effective use of technological and human innovation enablers will allow them to catch up with—and even surpass—their global competitors, some of whom have been optimizing and streamlining their processes for many years. As Xerox CEO Paul Allaire, whose firm has successfully created and marketed products of world-class quality, observed: "We're never going to out-discipline the Japanese on quality. To win, we need to find ways to capture the creative and innovative spirit of the American worker. That's the real organizational challenge."[9]

WHAT IS A PROCESS?

Adopting a process view of the business—a key aspect of process innovation—represents a revolutionary change in perspective: it amounts to turning the organization on its head, or at least on its side. A process orientation to business involves elements of structure, focus, measurement, ownership, and customers. In definitional terms, a process is simply a structured, measured set of activities designed to produce a specified output for a particular customer or market. It implies a strong emphasis on *how* work is done within an organization, in contrast to a product focus's emphasis on *what*.

A process is thus a specific ordering of work activities across time and place, with a beginning, an end, and clearly identified inputs and outputs: a structure for action. This structural element of processes is key to achieving the benefits of process innovation. Unless designers or participants can agree on the way work is and should be structured, it will be very difficult to systematically improve, or effect innovation in, that work.

Process structure can be distinguished from more hierarchical and vertical versions of structure. Whereas an organization's hierarchical structure is typically a slice-in-time view of responsibilities and reporting relationships, its process structure is a dynamic view of how the organization delivers value. Furthermore, while we cannot measure or improve hierarchical structure in any absolute sense, processes have cost, time, output quality, and customer satisfaction. When we reduce cost or increase customer satisfaction, we have bettered the process itself.

Some managers view the dynamic nature of processes in a negative, bureaucratic sense: "We can't do anything around here unless we follow a process." To the contrary, this book is based on the assumption that following a structured process is generally a good thing, and that there is nothing inherently slow or inefficient about acting along process lines.

A process approach to business also implies a relatively heavy emphasis on improving how work is done, in contrast to a focus on which specific products or services are delivered to customers. Successful organizations must, of course, both offer quality products or services and employ effective, efficient processes for producing and selling them. But U.S. companies spend twice as much researching and developing new products as new processes (these proportions are reversed in Japan[10]), and almost all of the amount spent on processes is targeted at engineering and manufacturing. Marketing, selling, and administrative processes receive very little investment. Adopting a process perspective means creating a balance between product and process investments, with attention to work activities on and off the shop floor.

Researchers of innovation in organizations frequently distinguish between product and process innovation, the former receiving both more attention by firms and more study by researchers. But at least one recent study has pointed out that the two types of innovation often seem to occur together.[11] Indeed, in service industries it is nearly impossible to distinguish between innovative new services offered to customers and the innovative processes that enable them. As these industries mature and seek innovation in their offerings, it is perhaps inevitable that they will adopt more of a process orientation.

Processes that are clearly structured are amenable to measurement in a variety of dimensions. Such processes can be measured in terms of the time and cost associated with their execution.

Their outputs and inputs can be assessed in terms of usefulness, consistency, variability, freedom from defects, and numerous other factors. These measures become the criteria for assessing the worth of the innovation initiative and for establishing ongoing improvement programs.

Taking a process approach implies adopting the customer's point of view. Processes are the structure by which an organization does what is necessary to produce value for its customers. Consequently, an important measure of a process is customer satisfaction with the output of the process. Because they are the final arbiters of process design and ongoing performance, customers should be represented throughout all phases of a process management program.

Processes also need clearly defined owners to be responsible for design and execution and for ensuring that customer needs are met. The difficulty in defining ownership, of course, is that processes seldom follow existing boundaries of organizational power and authority. Process ownership must be seen as an additional or alternative dimension of the formal organizational structure that, during periods of radical process change, takes precedence over other dimensions of structure. Otherwise, process owners will not have the power or legitimacy needed to implement process designs that violate organizational charts and norms describing "the way we do things around here."

Our definition of process can be applied to both large and small processes—to the entire set of activities that serves customers, or only to answering a letter of complaint. The larger the process, however, the greater the potential for radical benefit. A key aspect of process innovation is the focus on broad, inclusive processes. Most companies, even very large and complex ones, can be broken down into fewer than 20 major processes. IBM, for example, has identified 18 processes, Ameritech 15, Xerox 14,[12] and Dow Chemical 9. Key generic business processes include product development, customer order fulfillment, and financial asset management. Figure 1-1 depicts a typical set of broad processes for a manufacturing firm.

Because a process perspective implies a horizontal view of the business that cuts across the organization, with product inputs at the beginning and outputs and customers at the end, adopting a process-oriented structure generally means deemphasizing the functional structure of the business. Today almost every large

Figure 1-1   Typical Processes in Manufacturing
Firms

Operational
Product development
Customer acquisition
Customer requirements identification
Manufacturing
Integrated logistics
Order management
Post-sales service
Management
Performance monitoring
Information management
Asset management
Human resource management
Planning and resource allocation

business organization is characterized by the sequential movement of products and services across business functions—engineering, marketing, manufacturing, sales, customer service, and
so forth. Not only is this approach expensive and time-consuming,
it often does not serve customers well. In functionally oriented
organizations, handoffs between functions are frequently uncoor-
dinated. As a result, no one may be responsible for measuring or
managing the time and cost required to move products from laboratory to market or from customer order to receipt. Process innovation demands that interfaces between functional or product
units be either improved or eliminated and that, where possible,
sequential flows across functions be made parallel through rapid
and broad movement of information.

Major processes such as product development include activities that draw on multiple functional skills. New product designs
are generated by research and development, tested for market
acceptance by marketing, and evaluated for manufacturability
by engineering or manufacturing (see Figure 1-2). Processes that
involve order management and service cross the external boundaries of organizations, extending into suppliers and customers.
Consequently, viewing the organization in terms of processes and
adopting process innovations inevitably entails cross-functional
and cross-organizational change.

Adopting a process view implies a commitment to process

Figure 1-2 A Typical Cross-Functional Process

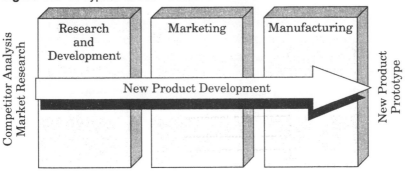

betterment. Much recent attention has been given to time reduction as an objective of business change;[13] cost reduction and quality improvement are also familiar. But no single objective is sufficiently ambitious for a process innovation initiative. Many firms have found that they can achieve multiple objectives with each process innovation initiative. Indeed, they must: customers demand cycle time reduction and output quality improvements, while the competitive and financial environments simultaneously demand that process costs be substantially reduced.

Some of these improvement objectives can be found within a functional (i.e., nonprocess) context. Manufacturing, for example, has been improving cycle time and quality for many years. But these improvements often are not perceived by the customer because of poor coordination with other functions. A product is manufactured more quickly, for example, but sits in the warehouse awaiting a customer credit check or resolution of a discrepancy in an order. Consequently, the impact of functional betterment, even when fully achieved, may be limited. Process improvement, on the other hand, whether internal or external, should immediately benefit the customer.

To some, a process orientation might imply a process industry, that is, an industry, such as chemicals, that produces a product continuously rather than in discrete units. But process improvement and innovation as we discuss them here apply to all industries. It may be easier to apply process thinking to manufacturing firms (both discrete and process), because structure and measurement have traditionally been applied to manufacturing processes, but the benefits of process thinking are clearly attainable by service industries as well.

WHAT CONSTITUTES INNOVATION?

Defined simply, innovation is, of course, the introduction of something new. We presume that the purpose of introducing something new into a process is to bring about major, radical change. Process innovation combines a structure for doing work with an orientation to visible and dramatic results. It involves stepping back from a process to inquire into its overall business objective, and then effecting creative and radical change to realize order-of-magnitude improvements in the way that objective is accomplished.

Process innovation can be distinguished from *process improvement,* which seeks a lower level of change. If process innovation means performing a work activity in a radically new way, process improvement involves performing the same business process with slightly increased efficiency or effectiveness. The actual level of benefit derived from operational betterment initiatives falls, of course, across a continuum, but in practice most firms seek either incremental or radical change. It is possible that process innovation might yield only incremental benefit, in which case we would classify it as an improvement.[14] We are also familiar with at least one instance in which a process improvement initiative yielded radical benefit, albeit in a narrowly defined process.[15]

For example, a firm that analyzes its customer order-fulfillment process and then eliminates redundant or non-value-adding steps is practicing process improvement. This activity might eliminate several unnecessary jobs, improve customer satisfaction, and reduce delivery time from three weeks to two. Another firm that looked at its order fulfillment process with an eye toward process innovation might provide customers with order entry terminals, eliminate its direct sales force, guarantee order fulfillment, arrange for a third party to manage its warehouse, and empower frontline personnel to handle all financial and shipping details. The latter firm might halve its costs and order fulfillment times. Even the best American performers in terms of quality (i.e., the highest scores in the Baldrige award competition) improve reliability, cycle time, inventory turns, and so forth an average of only 5% to 12%;[16] firms undertaking process innovation could not afford to be satisfied with these results.

There are other important differences between process

improvement and process innovation, among them, the locus of participation in organizational change, the importance of process stabilization and statistical measurement, the enablers and nature of change, and the degree of organizational risk. These are summarized in Figure 1-3.

Process innovation initiatives start with a relatively clean slate, rather than from the existing process. The fundamental business objectives for the process may be predetermined, but the means of accomplishing them is not. Designers of the new process must ask themselves, "Regardless of how we have accomplished this objective in the past, what is the best possible way to do it now?"

Whereas process improvement initiatives are often continuous in frequency, the goal being ongoing and simultaneous improvement across multiple processes, it is difficult to conceive of continuous process innovation. Process innovation is generally a discrete initiative. As suggested below, it is best combined with improvement programs, both concurrently across different processes and in a cycle of alternation for a single process.

Process improvement can begin soon after changes in a process are identified, and incremental benefits can be achieved within

Figure 1-3   Process Improvement versus Process Innovation

	Improvement	*Innovation*
Level of Change	Incremental	Radical
Starting Point	Existing process	Clean slate
Frequency of Change	One-time/continuous	One-time
Time Required	Short	Long
Participation	Bottom-up	Top-down
Typical Scope	Narrow, within functions	Broad, cross-functional
Risk	Moderate	High
Primary Enabler	Statistical control	Information technology
Type of Change	Cultural	Cultural/structural

months. Because of the magnitude of organizational change involved, process innovation often takes a much longer time. We know of no large organization that has fully identified and implemented a major process innovation in less than two years. Ford's widely reported elimination of three-quarters of its accounts payable staff, achieved by paying on receipt of goods and eliminating invoices, took five years from design to full implementation.[17] This implementation cycle was long despite the fact that the process Ford adopted was already operational within Mazda, its Japanese subsidiary, and was a relatively narrow process. In this situation, managing the changes required in accounts payable took longer than it should have, but process innovation champions should be prepared for a change cycle measured in years. Clearly, if the cycle of process innovation takes too long, the average improvement per year may not exceed that of process improvement, and for this reason we urge both haste and concurrent reliance on process improvement approaches aimed at realizing short-term gains in existing processes.

Bottom-up participation is a hallmark of continuous quality improvement programs;[18] employees are urged to examine and recommend changes in the work processes in which they participate. Visitors to quality-oriented companies frequently see banners, posters, and newsletters bearing quality-related slogans and progress measurements. Some quality-driven firms boast that every member of the organization is a member of a quality circle.

Process innovation is typically much more top down, requiring strong direction from senior management. Because large firms' structures do not reflect their cross-functional processes, only those in positions overlooking multiple functions may be able to see opportunities for innovation. A clerk in the shipping department is unlikely to conceive a radical redesign of the entire order-management process. Ideas from workers and lower-level managers, though they should be solicited, are likely to target incremental improvement. But inasmuch as workers and lower/middle managers are as likely to resist as to propose major change, implementers of process innovation must strive to gain commitment and buy-in at all levels of the organization. Encouraging participation in the process design activity can certainly facilitate this.

Process improvement programs, including those initiated under the quality banner, are generally applied to existing organizational structures and thus involve change in narrowly defined functional or subfunctional processes. Process innovation, on the other hand, involves definition of and innovation in broad, cross-functional processes. Just the identification of such processes, often for the first time, can lead to innovative ways of structuring work.[19]

Both process improvement and process innovation require cultural change. The necessary focus on operational performance, measurement of results, and empowerment of employees are all aspects of the cultural shift. But whereas it is possible to implement a continuous improvement program without making major changes in organizational structure—indeed, avoiding uncontrolled change is prerequisite to continuous improvement—process innovation involves massive change, not only in process flows and the culture surrounding them, but also in organizational power and controls, skill requirements, reporting relationships, and management practices. The wrenching nature of this organizational change is the most difficult aspect of process innovation and accounts at least partially for its typically long cycle times.

The primary enabler of continuous process improvement programs is statistical process control, a technique for explaining and minimizing sources of variation. Neither this, nor the other key quality techniques, is well adapted to the large variations in process outcomes produced by process innovation.

Process innovation also implies the use of specific change tools. One of these, information technology, has been hailed by many as the most powerful tool for changing business to emerge in the twentieth century. The dramatic capabilities of computers and communications are powerful enablers of process innovation; but though they have yielded impressive benefits for many firms, they have not been as fully exploited as they might be.

Human and organizational development approaches such as greater employee empowerment, reliance on autonomous teams, and flattened organizational structures are as key to enabling process change as any technical tool. In fact, information technology is rarely effective without simultaneous human innovations.

If continuous improvement involves relatively little reward, it also involves little risk. With a few exceptions (such as at Florida Power and Light, where an award-winning quality initiative was abruptly discontinued by a new CEO), quality programs do not die in a visible and obvious way; they are much more likely to fade into oblivion. Because process innovation initiatives have (or should have) well-defined and ambitious change objectives, failure to achieve these objectives, while not beyond coverup, is usually highly apparent. The level of change involved and the cross-functional nature of process innovation greatly heighten the risk of failure. Managers likely to lose people, power, or other resources as a result of process innovation will naturally be tempted to resist the change.

However different in character, continuous process improvement and process innovation present similar challenges. Both require a strong cultural commitment and high degree of organizational discipline, a process approach, a measurement orientation, and a willingness to change. A company that is unsuccessful at one will probably not succeed at the other.

But although continuous quality improvement may be good practice, it is not a prerequisite for success at process innovation. The skills and enablers of change are different. Success at quality initiatives is a qualification for success in process innovation, but so are many other types of corporate change that involve seriousness of purpose and flexibility—success in merging large organizations, success in downsizing, success in major product or service line innovations.

In practice, most firms need to combine process improvement and process innovation in an ongoing quality program (see Figure 1-4). Ideally (though not necessarily), a company will attempt to stabilize a process and begin continuous improvement, then strive for process innovation. Lest it slide back down the slippery slope of process degradation, a firm should then pursue a program

Figure 1-4   Process Improvement and Process Innovation

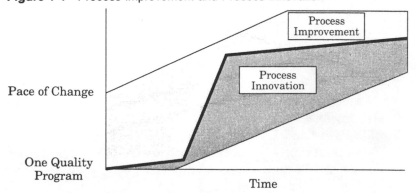

of continuous improvement for the post-innovation process. Furthermore, across an entire organization, innovation initiatives will be appropriate for some processes, continuous improvement initiatives for others.

In order not to confuse the employees who participate in, or are affected by, innovation and improvement initiatives, all such activities should be carried out within the context of a single quality program, and it should be clear which type of process change is under way for a given process at a particular time. For example, we found a very confusing situation at an electrical utility in which both improvement and innovation initiatives were being undertaken at the same time by different sponsors. The quality organization had begun continuous improvement programs along functional lines; the information systems function was sponsoring process innovation programs with radical change objectives, structured along major cross-functional process lines. Even the leaders of these two initiatives were unsure how they related to one another.

The differences between process improvement and innovation can make it difficult to combine the two. One way to facilitate their combination is to assign them to different managers. At Ameritech, where this was done, the manager with responsibility for process innovation and the manager assigned to quality and process improvement communicate regularly and the level of co-operation is high, perhaps because they report to the same person.

A firm should be aware that the risks of process innovation are at least proportional to the rewards. Given this equation, organizations that can avoid such wrenching change should probably do so. In environments that are relatively noncompetitive or in which basic business practices are not in question (e.g., some segments of the utility industry or other highly regulated businesses, or well-funded government organizations), continuous improvement may be preferred over process innovation. But competitive and other pressures force most firms to seek radical change.

PROCESS INNOVATION AND THE VOICE OF THE CUSTOMER

A correctly designed business process has the voice and perspective of the customer "built in." A process should be designed to produce outputs that satisfy the requirements of the customer.

A process innovation initiative should begin with a good understanding of who the customers of the process are and what they want out of it. Asking customers about their requirements and getting them to choose between process trade-offs should be explicit tasks. Improvement or innovation objectives should be primarily those of the customer.

In fact, the customer's perspective should be built not only into the final process design, but also into the early visioning and post-implementation activities. When possible, customers should be included on process design teams and, after a process design has been created, should participate in prototypes of the process and help refine the design. Ongoing measures of the process should be from the customer's perspective, with customers assessing process performance to the extent possible.

Processes such as order management and customer service extend across firm boundaries into the customer organization; in these processes, the customer is not a guest in the design activities, but an owner of them. An interorganizational process should be jointly designed and managed by the organizations whose boundaries it crosses. Costs or bottlenecks should not be passed from one firm to the other, but designed out of the process altogether. This change to a more "networked" view of processes is already beginning to have important consequences for day-to-day management and organization.[20]

ENABLERS OF PROCESS INNOVATION

The idea of focusing on enablers of innovation as potential drivers of change is perhaps radical. Conventional wisdom about business initiatives holds that they must first be planned in the abstract, without reference to specific tools or levers of change. In the traditional planning approach, we define "ends" (corporate objectives), the "ways" in which these ends are to be achieved (specific visions or critical success factors), and the "means" by which we expect to bring these ends about (enablers of change). Hayes, observing that abstract planning only rarely anticipates how a firm will address its environment with specific initiatives, argues that often the appropriate order is "means, ways, ends."[21] Provide the tools for change (including financial, technological, and human resources), and the specific directions in which to apply them will become apparent as the environment changes.

Although fresh and appealing, Hayes's view may not be entirely applicable to process innovation. Some sort of context for process innovation—a sense of the "end"—is necessary to provide focus and inspiration. Change enablers for process innovation are clearly "means" in Hayes's terminology. Rather than following in some fixed order, ends and means are better viewed as influencing each other. The use of information technology in processes, for example, can strongly influence, and should be influenced by, strategy. Regardless of the order of ends, ways, and means, it is important to consider the means or enablers of process innovation before a process design is solidified.

By virtue of its power and popularity, no single business resource is better positioned than information technology to bring about radical improvement in business processes. It is the least familiar of key resources, having existed in a form useful to businesses for a mere 40 years. Especially in hardware and communications components, IT capabilities have grown faster than those of other resources or technologies. We are only beginning to understand the power of information technology in business. But even as we begin to grasp existing IT capabilities, innovations make our perspectives obsolete.

Its great potential notwithstanding, IT cannot change processes by itself, nor is it the only powerful resource. There is a well-developed literature suggesting that the primary enablers of change in organizations are technology (including technologies not based on information) and organizational/human factors.[22] Process innovation can seldom be achieved in the absence of a carefully considered combination of both technical and human enablers.[23]

Possibilities for applying organizational and human factors to process innovation extend over a broad range. Throughout this book, we treat them as equal partners with information technology in effecting process change. For every example of IT as an enabler of new process designs, there is almost invariably an accompanying change in the organizational or human resource type. For example, creating a more empowered and diversified work force, eliminating levels of hierarchy, creating self-managing work teams, combining jobs and assigning broader responsibilities, and upgrading skills are some organizational and human resource changes that frequently accompany the use of IT.

Other process innovation resources—new approaches to

assessing financial costs and benefits,[24] and global business management, for example[25]—have been in managers' portfolios much longer than information technology and new organizational forms and thus may have less potential as enablers of radical innovation. Moreover, even these traditional resources work best when used in concert with information technology and human/organizational innovations.

Firms undertaking process innovation initiatives should not, of course, become preoccupied with particular enablers or tools. The objective of such efforts should be improved business performance, not a lesson in taking advantage of a new technology or alternative human resource tactic. But it is just as foolish to invent new process designs without examining firms that have successfully employed IT and human enablers in similar process innovation initiatives. Just as an architect brings to the design of a building knowledge of the technologies needed to operate it (e.g., elevators, air conditioning, plumbing, and so forth) and the types of people who will work in it, so a process designer must be cognizant of the technologies and people involved in making a process work.

OVERVIEW AND BACKGROUND OF THE BOOK

This book is the result of more than four years of research into process innovation, in both academic and consulting contexts. It reflects hundreds of conversations with executives and professionals in more than 50 companies. (These companies are listed in Appendix A.) In addition, we reviewed a wealth of case materials on other firms that were in the midst of process innovation. As does most useful business research, our work on process innovation describes the state of the art, and then attempts to go beyond it in terms of analysis, development of frameworks, and prescription.

The breadth of the research notwithstanding, process innovation remains more art than science. Many of the process innovation initiatives we studied had only recently begun. The approaches and methods described here are not the only possible routes to success. This book is for the managers of companies now embarking upon such initiatives, who need information about the nature of process innovation and the experiences of early movers. In describing the efforts of pioneers, the book is itself pioneering. The book is organized in three parts. We first present the components of a general approach to, and context for, process innovation. These components, although they do not form a detailed methodology, must be aspects of a successful process innovation program. The components of the approach are depicted in graphic form in the introduction to Part I and are discussed in detail in the chapters that follow. Chapter 2 examines the selection of processes for innovation, Chapters 3, 4, and 5 the enablers of process innovation, including information technology, information, and organizational and human resource approaches. In Chapter 6 we discuss the creation of a vision for processes and its relation to corporate strategy, and in Chapter 7, the issue of measurement and short-term improvement of existing processes. Chapter 8 concludes Part I with a detailed review of the design and implementation of new processes.

Part II is devoted to implementation issues associated with radical process change. We recommend in Chapter 9 approaches to managing successfully the kind of large-scale organizational change that attends process innovation. In Chapter 10, we view the implementation of information systems within a process innovation context—that is, we examine how, given a new process design, we can employ IT to rapidly and effectively implement systems that support it.

In Part III, we focus on specific approaches that firms are employing to innovate key operational and managerial processes. Chapter 11 examines innovative approaches to product design, development, and manufacturing processes. Chapter 12 focuses on customer-facing processes, including marketing and sales management, order management, and customer service. Chapter 13 discusses innovation in management and administrative processes, which have been neglected even by firms undertaking process innovation elsewhere.

In Chapter 14, after summarizing key points of the book, we urge both caution and haste in adopting process innovation. Given the rapidly growing popularity of this approach to business improvement, a balanced view of risks and rewards is essential. But neither these cautions, nor the myriad complexities introduced throughout the book, should be permitted to diminish the appeal of this exciting approach to enhancing business performance.

Appendix B traces the historical roots of process innovation. This discussion further elucidates the distinctions between process innovation and earlier approaches to operational improvement. The current relationship of process innovation to quality, sociotechnical, work design, and competitive systems initiatives is also discussed.

Notes

1See Hirotaka Takeuchi and Ikujiro Nonaka, "The New New Product Development Game," *Harvard Business Review* (January–February 1986): 137–146. For comparisons between automotive firms, see Kim B. Clark and Takahiro Fujimoto, *Product Development Performance: Strategy, Organization, and Management in the World Auto Industry* (Boston: Harvard Business School Press, 1991).

2See, for example, Kaoru Ishikawa, *What Is Total Quality Control? The Japanese Way* (Englewood Cliffs, N.J.: Prentice-Hall, 1985): 171–184.

3James P. Womack, Daniel T. Jones, and Daniel Roos, *The Machine That Changed the World* (New York: Rawson Associates, 1990).

4Susan Helper, "How Much Has Really Changed Between U.S. Automakers and Their Suppliers?," *Sloan Management Review* (Summer 1991): 15–28.

5For a rigorous discussion of these types of processes in the retail industry, see James V. McGee, "Implementing Systems Across Boundaries: Dynamics of Information Technology and Integration," Ph.D. diss., Harvard Business School, 1991.

6This movement toward interdependence with reference to information technology has been described in John Rockart and James Short, "IT in the 1990's: Managing Organizational Interdependence," *Sloan Management Review* (Winter 1989): 7–18.

7Melissa Mead and Jane Linder, "Frito-Lay, Inc.: A Strategic Transition (A)," 9-1187-012. Boston: Harvard Business School, 1987; also Nicole Wishart and Lynda Applegate, "Frito-Lay, Inc.: HHC Project Follow-Up," 9-190-191. Boston: Harvard Business School, 1990.

8See, for example, John F. Krafcik, "Triumph of the Lean Production System," *Sloan Management Review* (Fall 1988): 41–52.

9Allaire is quoted in Brian Dumaine, "The Bureaucracy Busters," *Fortune*, June 17, 1991: 37.

10Edwin Mansfield, "Industrial R&D in Japan and the United States: A Comparative Study," *American Economic Review* 78 (1988): 223.

11Urs E. Gattiker, *Technology Management in Organizations* (Newbury Park, Calif.: Sage Publications, 1990): 19–20.

12The 14 processes identified by Xerox apply only to its document-processing business. Since they were originally identified, Xerox has added and changed some processes, and the firm is now developing a new process architecture.

13The most prominent example of this movement is George Stalk, Jr., and Thomas M. Hout, *Competing Against Time* (New York: Free Press, 1990).

14See, for example, R. D. Dewar and J. E. Dutton, "The Adoption of Radical and Incremental Innovations: An Empirical Analysis," *Management Science* 32:11 (1986): 1422–1433.

15Thomas H. Davenport, "Rank Xerox U.K. (A) and (B)" N9-192-071 and N9-192-072. Boston: Harvard Business School, 1992.

16U.S. Government Accounting Office, "Management Practices: U.S. Companies Improve Performance Through Quality Efforts," Report to the Honorable Donald Ritter, House of Representatives, May 1991.

[17]The Ford example is described in Thomas H. Davenport and James E. Short, "The New Industrial Engineering: Information Technology and Business Process Redesign," *Sloan Management Review* (Summer 1990): 11–27; also Michael Hammer, "Reengineering Work: Don't Automate, Obliterate," *Harvard Business Review* (July–August 1990): 104–112.

[18]Dean M. Schroeder and Alan G. Robinson, "America's Most Successful Export to Japan: Continuous Improvement Programs," *Sloan Management Review* (Spring 1991): 67–81.

[19]This aspect of the difference between improvement and innovation programs is discussed in Robert B. Kaplan and Laura Murdock, "Core Process Redesign," *McKinsey Quarterly* (Summer 1991): 27–43.

[20]Raymond E. Miles and Charles C. Snow, "Organizations: New Concepts for New Forms," *California Management Review* 18:3 (Spring 1986): 62–73.

[21]Robert Hayes, "Strategic Planning: Forward in Reverse?" *Harvard Business Review* (November–December 1985): 111–119.

[22]For a review of this literature from the process perspective, see Dorothy Leonard-Barton, "The Role of Process Innovation and Adaptation in Attaining Strategic Technological Capability," 1991–007. Boston: Harvard Business School, Division of Research, 1990; republished in the *International Journal of Technology Management* 6 (1991): 3–4. For a review of the sociotechnical literature, see Kenyon B. DeGreene, *Sociotechnical Systems* (Englewood Cliffs, N. J.: Prentice-Hall, 1973).

[23]See Shoshana Zuboff, *In the Age of the Smart Machine: The Future of Work and Power* (New York: Basic Books, 1988) and Richard E. Walton, *Up and Running: Integrating Information Technology and the Organization* (Boston: Harvard Business School Press, 1989).

[24]On the cost side, see H. Thomas Johnson and Robert S. Kaplan, *Relevance Lost: The Rise and Fall of Management Accounting* (Boston: Harvard Business School Press, 1987); on the benefit side, Tom Copeland, Tim Koller, and Jack Murrin, *Valuation: Measuring and Managing the Value of Companies* (New York: John Wiley, 1990).

[25]Christopher A. Bartlett and Sumantra Ghoshal, *Managing Across Borders: The Transnational Solution* (Boston: Harvard Business School Press, 1989).

Part I

A Framework For Process Innovation

Introduction

Hundreds of firms throughout Europe and America are undertaking process innovation initiatives. Close examination of a number of early initiatives has led us to create a high-level framework to guide process innovation. We present the primary components of a process innovation approach over several chapters. The purpose of these chapters is not to provide a detailed methodology, inasmuch as specific activities undertaken by specific firms will be variations upon these components, differing in order, emphasis, and flavor. Rather, we believe that each component is necessary in some form for a successful innovation initiative. We know of no completed radical process change that has not involved all of these components in some form, whether implicit or explicit, and we are aware of several failed efforts that did not employ all these steps.

Process innovation initiatives are inherently distinct from business as usual. In fact, a number of researchers have argued that the existing Western management paradigm views both improvement and innovation as lying outside routine management activities.[1] Our experience suggests that companies can institutionalize incremental improvement through organizational and cultural change programs, with those doing the work identifying and implementing small changes in product and process. But we see no realistic way to conduct process innovation during the normal course of business. Companies typically treat innovation activities as special tasks, assigned to project teams or task forces. We believe that the project or special initiative structure is the only way to accomplish radical innovation. A project orientation appears to be the best way to introduce and gain experience with process thinking and innovation within an organization, and only ad hoc, cross-functional teams may be able to innovate processes that traverse organizational boundaries and areas of management responsibility.

The matrix in Figure I-1 categorizes various process-based operational improvement methods in terms of the relationship between level of change and context or frequency of application. Value analysis approaches such as process, overhead, and activity value analysis are project-oriented, but yield only incremental improvement. Quality-oriented methods such as business process improvement and activity-based costing mechanisms are intended to yield continuous but incremental improvement. These approaches are treated in greater detail in Chapter 7, where we discuss improvement of the existing process in the overall context of process innovation.

Only process innovation is intended to achieve radical business improvement. It is a discrete initiative that must be combined with other initiatives for ongoing change. The notion of continuous innovation advanced in the management literature pertains to product, rather than process innovation;[2] we have not observed continuous process innovation of the type and scale seen in a typical one-time process innovation effort and believe that such levels of innovation would be difficult to maintain and coordinate on a continuous basis. If nothing else, people and organizations need periods of rest and stability between successive innovation initiatives.

Although it may not be possible to achieve radical innovation

Figure I-1 Approaches to Business Improvement

Context \\ Outcome	Project / One-Time	Continuous Improvement / Ongoing
Incremental Improvement	• Activity value analysis • Overhead value analysis • Process value analysis	• Total quality management • Business process improvement • Activity-based costing
Radical Innovation	Process innovation (reengineering, business process redesign)	Not meaningful

while practicing continuous improvement, companies need to learn how to do both—concurrently across different processes, and cyclically for a single process. Ultimately, a major challenge in process innovation is making a successful transition to a continuous improvement environment. A company that does not institute continuous improvement after implementing process innovation is likely to revert to old ways of doing business.

Our framework for process innovation consists of five steps: identifying processes for innovation, identifying change enablers, developing a business vision and process objectives, understanding and measuring existing processes, and designing and building a prototype of the new process and organization (Figure I-2). Excepting change enablers, each step is discussed in a chapter; each of the three major enablers of change merits its own chapter.

Although the sequence of the activities in Figure I-2 may vary, aspects of the ordering are important. Selecting processes for innovation, for example, should be done early in order to focus effort and resources.

Figure I-2 A High-Level Approach to Process Innovation

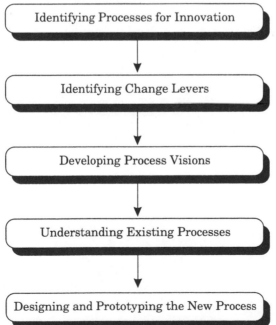

Our high-level approach also presumes the establishment of an infrastructure for process innovation. As with any project-oriented business initiative, a project team must be selected and trained. Aspects of team selection that influence the change management process are discussed in Chapter 10. The team infrastructure required for innovation, though similar in some respects to that required for organizationwide continuous improvement,[3] differs most significantly in the need for innovation teams to be familiar not only with particular processes, but also with the enablers of change discussed later.

Notes

[1]Richard D. Sanders, Gipsie B. Ranney, and Mary G. Leitnaker, "Continual Improvement: A Paradigm for Organizational Effectiveness," *Survey of Business* (Summer 1989): 12–20.

[2]Ikujiro Nonaka, "The Knowledge-Creating Company," *Harvard Business Review* (November–December 1991): 96–104.

[3]H. James Harrington, *Business Process Improvement: The Breakthrough Strategy for Total Quality, Productivity, and Competitiveness* (New York: McGraw-Hill, 1991).

Chapter 2

Selecting Processes for Innovation

Process innovation must begin with a survey of the process landscape to identify processes that are candidates for innovation. Both the overall listing of processes and the focus on those requiring immediate innovation initiatives are crucial to the success of innovation efforts. The selection process establishes the boundaries of the processes that are to be addressed, enabling a firm to focus on those most in need of radical change.

The principal activities in the selection process are listed in Figure 2-1. The first is to identify the major processes in the organization. An informed selection can be made only when all of the organization's processes are known. A survey also serves to determine process boundaries that help establish the scope of initiatives for individual processes.

ENUMERATE MAJOR PROCESSES

Considerable controversy revolves around the number of processes appropriate to a given organization. The difficulty derives from the fact that processes are almost infinitely divisible; the

Figure 2-1 Key Activities in Identifying Processes for Innovation

- Enumerate major processes
- Determine process boundaries
- Assess strategic relevance of each process
- Render high-level judgments of the "health" of each process
- Qualify the culture and politics of each process

activities involved in taking and fulfilling a customer order, for example, can be viewed as one process or hundreds. The "appropriate" number of processes has been pegged at from two to more than one hundred. The three major processes identified by Rockart and Short—developing new products, delivering products to customers, and managing customer relationships[1]—are themselves highly interdependent; and Harvard researchers working on order management issues have argued for only two processes (1) managing the product line, and (2) managing the order cycle.[2] A well-known management consulting firm has asserted that there are only three or four "core" processes, though not all business activities are part of these processes.[3] Finally, at least one firm, Xerox Corporation, has identified a larger number of processes, but has focused its change efforts on those it considers most critical or core. IBM, which in the 1980s had defined at least 140 processes across the corporation, is today working with 18 much broader processes.

The objective of process identification is key to making these definitions and determining their implications. If the objective is incremental improvement, it is sufficient to work with many narrowly defined processes, as the risk of failure is relatively low, particularly if those responsible for improving a process are also responsible for managing and executing it. But when the objective is radical process change, a process must be defined as broadly as possible. A key source of process benefit is improving handoffs between functions, which can occur only when processes are broadly defined. Moreover, if a process output is minor, radically changing the way it is produced is likely to result in suboptimization or, at best, only minor gains.

As noted earlier, most of the companies that have identified their processes in the context of process innovation have enumerated between 10 and 20. Key processes identified by IBM, Xerox, and British Telecom[4] are presented in Table 2-1. The appropriate number of processes reflects a trade-off between managing process interdependence and ensuring that process scope is manageable. The fewer and broader the processes, the greater the possibility of innovation through process integration, and the greater the problems of understanding, measuring, and changing the process.

Our experience leads us to set the appropriate number for major processes at between 10 and 20. Within this range—which

Table 2-1 Key Business Processes of Leading Companies

IBM	Xerox	British Telecom
Market information capture	Customer engagement	Direct business
Market selection	Inventory management and logistics	Plan business
Requirements	Product design and engineering	Develop processes
Development of hardware	Product maintenance	Manage process operation
Development of software	Technology management	Provide personnel support
Development of services	Production and operations management	Market products and services
Production	Market management	Provide customer service
Customer fulfillment	Supplier management	Manage products and services
Customer relationship	Information management	Provide consultancy services
Service	Business management	Plan the network
Customer feedback	Human resource management	Operate the network
Marketing	Leased and capital asset management	Provide support services
Solution integration	Legal	Manage information resource
Financial analysis	Financial management	Manage finance
Plan integration		Provide technical R&D
Accounting		
Human resources		
IT infrastructure		

leaves us with some cross-process activity, but renders each process small enough to be understood—change management is only very difficult, rather than impossible. Constricting the range also permits us to identify both operational and management processes and to find different approaches to redesigning each type. This does not mean that all of the identified processes will be of the same importance, or even that innovations will be identified for all of them.

Some firms may wish to maintain both broad and narrow processes. While addressing broad processes in an innovation context, they may also be working on narrow processes from an improvement context. In order to avoid confusion, some mapping of narrow processes to broad processes should be developed. That is, a broad process like order management should be broken down into its constituent processes (or subprocesses, if that is the preferred terminology). The mapping need not be perfect—a narrow process, for example, could cut across two broader processes—but it should provide some orientation for all participants in process management initiatives. Among the firms we have researched, only British Telecom has developed such a hierarchy of broad and narrow processes.

Whatever the number of processes identified, the identification process should be understood to be exploratory and iterative. As a process becomes the focus for innovation or improvement efforts, its boundaries and relative importance become much clearer. Most companies that have worked on their processes for a number of years have revised their original lists.

Certain strategic-planning and systems-planning approaches that have a business process perspective may be useful for identifying processes. For example, Porter's framework for organizing a company's business activities emphasizes the usefulness of identifying and exploiting linkages in the value chain as a means of identifying both competitive opportunities and high-level processes,[5] and Ives and Learmonth have developed a generic customer-resource life cycle that describes the process by which a customer acquires, uses, and disposes of resources as a vehicle for identifying strategic information systems.[6] Neither of these approaches describes explicitly how a company should define these frameworks for its own use.

In the context of business process improvement, Harrington suggests having executives jot down the processes for which they are responsible and analyzing and distilling their lists to arrive at the company's slate of processes. This top-down approach, guided by our rule of thumb of 10 to 20 major processes, can be fairly effective. A major manufacturing concern used a similar approach to define its key processes. A candidate set of processes was developed by the corporate process-innovation staff and external consultants and presented to senior executives, who refined and revised the process list during a facilitated workshop discussion.

DETERMINE PROCESS BOUNDARIES

Once the processes have been identified at a high level, the boundaries between those processes need to be managed. Because process definition is more art than science, boundaries are arbitrary. A number of questions may help to define boundaries, among them:

- When should the process owner's concern with the process begin and end?
- When should process customers' involvement begin and end?
- Where do subprocesses begin and end?
- Is the process fully embedded within another process?
- Are performance benefits likely to result from combining the process with other processes or subprocesses?

Inasmuch as the end of any process is the beginning of another, either within or outside the organization, process innovation will often result in changes to upstream and downstream processes. Consequently, process management is best viewed as an iterative activity, in which subsequent innovation in one process gives rise to a need to reinnovate, or at least modify, others.

ASSESS STRATEGIC RELEVANCE

Having identified the boundaries of its major processes, a company must select individual processes for innovation. Our experience suggests that the scope of the innovation effort should be based on an organization's capabilities and resources. IBM has innovation initiatives under way in all 18 of its key processes at once, but most companies cannot successfully deal with innovation on such a scale. Even given a clear need to redesign, most organizations would lack sufficient resources—people, funds, and time—to do so. Beyond resources, most organizations could not endure the magnitude of organizational change that innovating all processes simultaneously would precipitate. An organization must understand the level of change and upheaval it can endure and must use that knowledge to determine how many processes it can successfully innovate.

Moreover, simultaneous change in multiple processes can be difficult to coordinate. Because redesigned processes must still interface with upstream and downstream processes, process owners or managers must communicate frequently about directions and interface points. If changes in one process must be coordinated with changes in, say, 17 others, it will be very difficult to make much headway.

Nevertheless, some firms find that they must work on groups of processes to solve particular business problems. Xerox, for example, found that to affect the time it takes to bring products to market it had to address not only its product design and engineering processes, but also processes such as manufacturing, service, and logistics. Similarly, IBM discovered that to speed delivery of custom-built products to customers, it had to address the production, logistics, and customer fulfillment processes.

Most companies choose to address a small set of business processes in order to gain experience with innovation initiatives, and they focus their resources on the most critical processes. Each successful initiative becomes a model for future efforts. We have identified four criteria that might guide process selection: (1) the process's centrality to the execution of the firm's business strategy, (2) process health, (3) process qualification, and (4) manageable project scope. Ideally, all four factors should favor the selection of a particular process; in practice, results are often ambiguous, and differential weighting of the factors must be applied.

The most obvious approach to process selection is to select the processes most central to accomplishing the organization's strategy (this presumes that the organization has a well-articulated strategy). Many firms' strategies, for example, focus on improving relationships with customers. One aspect of a customer-focused strategy is the provision of superior customer service at every point at which customer and enterprise meet. Such companies are likely to select for innovation processes at the customer interface, such as order management and customer service. In a more formal approach, the associations between business goals and business processes are identified and some sort of prioritization is done, based, for example, on the number and importance of goals the process supports. This approach to process selection is part of a number of planning and improvement methods.[7]

Selection on the basis of health targets redesign processes that are currently problematic and in obvious need of improvement. Among the many symptoms of unhealthy processes is the existence of multiple buffers, reflected in work-in-process queuing up at each step. One study of idle versus processing time for work-in-process revealed that, for the average process, actual working time comprises only .05% to 5% of total elapsed time.[8] IBM Credit

Corporation discovered, for example, that value-added time for the financing-approval process amounted to only 90 minutes out of a three- to four-day cycle time. We have seen several insurance companies in which the time spent actively processing and underwriting an insurance policy was only two or three hours, in an overall process cycle of more than 20 days. Process health is also suspect if a process crosses many functions and involves many narrowly defined jobs or has no clear owner or customers. Good indicators here are if no one gets upset when the process product is late or over budget, or no one is sure whom to call when deficiencies are noted.

Process health assessment at Cigna is aided by a process innovation group created by the insurance company to assist business units in their process innovation efforts. Upon receiving a request from a business unit, the group conducts a brief study to evaluate opportunities for process improvement. The group then advises the business unit whether to proceed with the innovation.[9] Among the first processes Cigna selected for innovation were in reinsurance, a relatively small business in which it was losing ground. Believing that the division had no choice but to pursue radical change, the head of the business unit told her employees: "If we succeed, half of us will no longer have a job in this division. If we fail, none of us will."

Even after a specific process has been selected, the enterprise must define the process in a way that results in a manageable innovation project scope. Some of the common ways of defining the process are by geography or by product line. One manufacturer seeking to improve the product development process decided to focus on its most profitable product line. After achieving dramatic results for this product, the next step was to begin introducing the other products to the new process.

A final point regarding process selection relates to process "qualification," an activity we observed in nearly every company in which we observed process innovation. The primary goal of process qualification is to gauge the cultural and political climate of a target process. A process "consultant," a key role discussed in a later chapter, is engaged to select only processes that have a committed sponsor and exhibit a pressing business need for improvement. Where commitment is lukewarm or the business need less than dramatic, the consultant would be expected to advise against attempting process innovation.

The determination of high-priority processes is often fairly obvious. At Continental Bank, because the firm had focused heavily on building broad relationships with corporate customers, the relationship management process was a natural place to start. At Bethlehem Steel, the process of scheduling a customer order into the production schedule was chaotic and inefficient and often led to customer dissatisfaction and sales force frustration. Therefore, the customer service process, which included production scheduling, was an easy selection. When Federal Mogul's Chassis Products Division analyzed its business, executives concluded that the failure to deliver new product prototypes quickly was a major impediment to increased sales. Applying radical change to this process was then not a difficult decision.

Regardless of how much care is taken in the identification and selection of processes, the process landscape will almost inevitably change with time and experience. The firms we have studied spent several years traveling the process innovation path, in the course of which most changed the number of processes they defined, the boundaries of individual processes, and the relative priorities of processes as candidates for innovation initiatives.

THE FATE OF PROCESSES NOT SELECTED

Processes not selected for early innovation initiatives must still be addressed. Recognizing that it can only be approximate, since business directions often change quickly and the resources for process innovation may rise and fall unpredictably, we advise creating a timetable for initiating innovation efforts. We do not recommend extending such a timetable over more than three years, since beyond this period predictions of business priority and resources are likely to be poor. Finally, innovation initiatives should be undertaken soon after detailed process qualification activities, since management commitments and process capabilities are also volatile over a period of several years.

Another alternative, discussed in the previous chapter, is to apply process improvement to processes not selected for innovation. This poses two problems. One, process improvement relies on a completely different set of methods and approaches. And two, it is difficult to apply incremental improvement to processes scoped broadly for purposes of achieving radical change. Because the bottom-up, participative nature of process improvement presumes a process that workers at lower levels in the organization can get their arms around, it may be necessary to break up the process into less unwieldy subprocesses. Should it later become a candidate for an innovation initiative, the process can be attacked in its broader form.

At least one firm, IBM, is pursuing improvement efforts concurrently with process innovation within the same process. The innovation efforts focus on broad processes, the improvement efforts on narrower constituent subprocesses. But it is not clear to many lower-level IBM process-improvement teams how their efforts relate to the enterprise-level process innovation initiatives. Although the notion of conducting top-down innovation and bottom-up improvement at the same time might seem appealing, much of the process improvement activity is likely to be for naught, inasmuch as teams might strive to improve narrow processes that do not even exist in the new process design. Whenever these activities are to be attempted simultaneously, it should be made clear to all parties that the improvements are likely to be short-lived and should therefore be implemented quickly if they are to yield any benefit.

SUMMARY

Identifying and selecting processes for innovation is an important prerequisite to process change. But this step is also significant in its own right. Without some focus on critical processes, an organization's energies, resources, and time will be dissipated. Especially during the early phases of process innovation, it is important that an organization demonstrate some successes. It can do so only if it is selective in the processes it chooses to innovate.

After a process has been selected for innovation, a firm can begin to think about how it will create quantum improvements in the process and what change tools it will employ. The next three chapters address the issue of how radical change in process designs is enabled. Each considers a different enabler of process change. Chapter 3 begins this discussion with an in-depth analysis of the role played by information technology in process innovation. Chapters 4 and 5 address the enablers of information and human/organizational factors.

Notes

[1]John F. Rockart and James E. Short, "Information Technology and the New Organization: Towards More Effective Management of Interdependence," working paper CISR 180, MIT Sloan School of Management, Center for Information Systems Research, September 1988.

[2]Benson P. Shapiro, John J. Sviokla, and V. Kasturi Rangan, "It Fell Through the Cracks," 9-591-098. Boston: Harvard Business School, 1991; and Sviokla's presentation at Ernst & Young, October 1990.

[3]Robert B. Kaplan and Laura Murdock, "Core Process Redesign," *McKinsey Quarterly* 2 (Summer 1991): 27–43.

[4]British Telecom has identified multiple levels of processes. The 15 listed in Table 2-1 are at the second level; the first-level processes are: (1) Manage the Business, (2) Manage People and Work, (3) Serve the Customer, (4) Run the Network, and (5) Support the Business.

[5]Michael E. Porter, *Competitive Advantage: Creating and Sustaining Superior Performance* (New York: Free Press, 1985).

[6]Blake Ives and Gerard P. Learmonth, "The Information System as a Competitive Weapon," *Communications of the ACM* (December 1984): 1193–1201.

[7]Maurice Hardaker and Bryan K. Ward, "Getting Things Done," *Harvard Business Review* (November–December 1987): 112–120.

[8]George Stalk, Jr., and Thomas M. Hout, *Competing Against Time* (New York: Free Press, 1990).

[9]Marlene Ammarito, director, Agency & Financial Systems, Cigna Systems presentation at Management in the Information Technology Era, R&D Forum, Workshop on Organizational Transformation, Cambridge, Mass., October 22–23, 1990.

Chapter 3

Information Technology as an Enabler of Process Innovation

Because it is both extremely important and not well understood by many managers, we begin our discussion of specific enablers of process innovation with information technology. Since they first entered the business environment in the 1950s, computers have been closely tied to the way work is carried out. One might even argue that information technology began to radically alter work—its location, speed, quality, and other key characteristics—with the advent of the telephone. Computers and telephones clearly benefit the business processes of the companies that employ them. Telephones shrink time and distance, enabling firms, for example, to monitor field sales on a daily basis and act accordingly. Computers speed the pace of many work activities and, at the same time, drastically reduce the need for human labor. Home offices of insurance companies, for example, used to be vast halls of closely packed desks at which clerks with adding machines made actuarial or financial calculations. Today, those halls are filled by rows of mainframe computers and disk drives, and the clerks are gone.

But as Joanne Yates observes in her book on the impact of information technology from 1850 to 1920, such inventions as the telegraph, telephone, and even the vertical filing cabinet, though they were immediately applicable to business demands of the time, did not immediately alter business practice.[1] In several case studies of how companies adopted these technologies, Yates found multidecade lags between early adoption and the point at which the technologies led to significant changes in the organization's "system" or processes.

Similarly, the computer's potential for changing processes (or procedures, as they were likely to be called at the time) was widely acknowledged early on. A shoe company executive, for example, described the introduction of a computer in his company between 1954 and 1958 as a major process breakthrough, citing inventory reductions from 11 to 8 million pairs of shoes and reductions in cycle time from 15 to 4 days, all with greatly increased product complexity and better service.[2] "A philosophy dedicated to the status quo will never put the most sophisticated tools of the modern world at the service of industry,"[3] one writer on the application of computers observed more than thirty years ago. Moreover, the systems analyst, who had responsibility for designing new computer applications, was expected to redesign processes first. Even newspaper help-wanted advertisements from the late 1960s noted that systems analysts were first to recommend changes in procedures, and then to apply computer applications;[4] a typical description of the analyst's role from twenty years ago included, for example, the "study and analysis of operations performed by qualified individuals in factories and offices."[5]

But although published sources do not say so, informal discussions with information systems and business executives who worked in the early days of computers suggest that little such process or procedure change was carried out. What did occur was largely incremental change. There are several possible reasons for this failure to innovate. One, it is unlikely that systems analysts were empowered by their employers to make—or even recommend—fundamental changes in procedures. Two, user executives probably devoted too little time to understanding the function of a system in relation to the function of the business— a problem we still see in many business environments. And three, there were no methodologies or formal approaches, or even structured lists of ideas, for using IT to bring about process or procedure change.

The desire to build information systems across functional boundaries is also not new. "A further disadvantage of being preoccupied with organization charts is that it becomes difficult to cut across departmental lines in developing the kind of system which ties together different departments and divisions," observed a manager at General Electric in 1960. "I, for example, am located in the production area. The systems that our group has been developing reach out and deal with marketing, financial information, costing, and, in fact, the general management problems of the entire business."[6] That many managers continue to argue for cross-functional systems suggests that the development of such systems never flourished or that it ended at some point. Cross-functional systems have probably always involved complicated issues of organizational politics and differing functional needs for information.

A handful of major applications initiatives of the 1960s, 1970s, and early 1980s are evidence of substantial cross-functional process changes. Many of the fabled "strategic systems" of that period were successful precisely because they enabled or led to major business process changes. Airline reservation systems radically simplified internal reservations processes, then changed the interorganizational processes involving travel agents.[7] These systems were and still are highly cross-functional, involving marketing, sales, and many aspects of operations. American Hospital Supply's customer-premise ordering system changed that firm's processes for managing customer inventories (somewhat consciously, as noted in a Harvard Business School case study[8]), and the McKesson system for drugstore replenishment has similar benefits for both internal and customer-oriented processes.[9] All these systems are notable for introducing innovation in customer-oriented processes, with subsequent rapid revenue, profitability, or share growth. The most successful users of information technology appear to have created process innovation without necessarily being aware of it.

On the quality side, there has, until recently, been little mention of information technology's role in process improvement. None of the classic quality texts singles out information technology as an important process improvement resource. Even a mid-1980s document on process management from IBM, which is arguably the world's preeminent information technology provider, neglects to mention the role of IT in process improvement.[10] (The company has since discovered the importance of this relationship and advanced it considerably, as we note throughout this book.)

Only very recently have quality-oriented process improvement experts begun to speak of the role of IT. Harrington, in perhaps the first book to address business process improvement directly,[11] refers to "automation" as an important tool in process improvement. Although his work is useful in describing how companies can begin to take a process improvement view of their businesses, there is little discussion of the impact of IT; indeed, Harrington's use of the term "automation" implies little role for technology in process change. Even more recent discussions of process improvement fail to mention information technology.[12]

Our research and consulting work has turned up another aspect of the history of process thinking that bears analysis. We have found in many companies that key processes were last "designed" (to the degree that they were designed at all) well before the rise of information technology. There could have been little consideration of process enablement through information technology when Procter & Gamble, for example, developed its fabled brand management system in the late 1940s. Over the past several years, the company has carried out a redesign of the process that shifts to the management of brand categories. The redesign has employed category management teams comprising IT specialists to take better advantage of IT and information in distribution processes.

Yates's work on early technologies suggests that it may take many years for companies to fully embed a technological advance in their organizations. We have had sophisticated commercial computers and data communications at our disposal for approximately 35 years. In terms of price and functionality, these technologies have become incredibly useful and usable. It is time to capitalize on them fully by employing them as enablers for business process innovation.

INFORMATION TECHNOLOGY AND THE PRODUCTIVITY CRISIS

Macroeconomic productivity figures suggest that firms have not often employed IT for significant change, despite radical improvements in IT functionally; over the same telephone lines that once carried only voices and static today travel purchase orders, large amounts of money, product design blueprints, marketing brochures, and meetings and conferences. The computer, which initially automated calculations, today advises decision makers and even makes decisions, collects and makes accessible vast stores of text, numbers, and graphic images, simulates a great variety of processes and environments (including limited but growing aspects of "reality"), and monitors and controls the performance of devices ranging from spaceships to artificial hearts.

The power and potential of computing has been matched by its penetration of the workplace and even the home. In the U.S.

business environment between 1978 and 1985, IT more than tripled as a share of total capital equipment stock, growing from 1.8% to 7.8%.[13] In early 1988, information technology capital accounted for 42% of total business equipment expenditures in the United States,[14] and by 1989, with computers in almost 14 million households, some 75 million people were using a computer at home, work, or school.[15]

But these outlays and penetration notwithstanding, at a macroeconomic level information technology has fallen short of its promise for effecting business transformation. In the foreword to a volume summarizing a major study on the business impact of information technology at MIT, Lester Thurow observes, "Specific cases in which the new technologies have permitted huge increases in output or decreases in costs can be cited, but when it comes to the bottom line there is no clear evidence that these new technologies have raised productivity (the ultimate determinant of our standard of living) or profitability. In fact, precisely the opposite is true. There is evidence, in the United States at least, that the investment in the new technologies has coincided with lowered overall productivity and profitability."[16]

The most aggressively pessimistic conclusions about IT and productivity have been advanced by Steven Roach. In a series of economic analyses using U.S. government productivity data, Roach argues that massive increases in spending on IT have yielded insufficient gains in productivity. He is particularly critical of IT spending in the service sector, where IT and overall capital spending are highest (the service sector owns more than 85% of the installed base of IT)[17] and productivity increases lowest (an average of 0.8% per year since 1982). "I dispute the idea that we need ever-increasing computational power to get work done by white-collar workers in the service sector," Roach writes. "American business managers are hooked on technology."[18]

This disappointing conclusion has also been reached in manufacturing industries and in white-collar work more broadly. Loveman's econometric study of the effect of IT on productivity in manufacturing (in which productivity gains overall have been higher than in services) found no significant positive productivity impact from IT investments,[19] the same conclusion reached by a study of white-collar productivity gains from IT across all sectors.[20] More detailed studies of clerical jobs and their relationship to IT indicate that, although IT has eliminated some office jobs, the shift to a computerized office has also created many new jobs.[21] In the insurance industry, for example, although computers have eliminated some types of clerical jobs and significantly thinned the ranks of clerks generally, between 1961 and 1976, computer-related jobs in programming and systems analysis and operations expanded greatly.[22]

All of the studies reported above employed U.S. government productivity data, the validity of which depends on the ability to accurately measure industry, sector, or economywide inputs and outputs. But as Panko points out, output measurement is highly suspect at all levels, and is measured by the government in such a way as to understate productivity.[23] Nevertheless, although he questions assertions of an IT productivity crisis, Panko acknowledges that there is little positive evidence of major benefits from IT.

Conclusions drawn from the handful of nonproductivity-oriented studies are no more sanguine. Diogo Teixeira's research in the banking industry, for example, suggests that although customers often benefit from IT investments, the investing firms only raise the entire industry's costs while creating excess capacity and lowering profitability.[24] Paul Strassman, in a broad analysis of the relationship between IT and business returns, opens his book with the observation that "there is no relationship between expenses for computers and business profitability."[25]

At the level of the company or business unit, most of the research on IT investment and business performance is highly anecdotal or case-study-based, and the analysis is rarely rigorous. As one review noted, "The data rarely provide a more detailed description of how or if the innovation created changes in the organization's underlying work processes or the structure of an individual worker's job."[26]

Although there are no certainties about the overall impact of IT on business economics, there are numerous large-scale examples of IT investment with little or no associated process change. The most prevalent use of computers by individuals in business is word processing—hardly a process innovation.[27] A study of the role of technology in office work with a detailed focus on word processing found that most implementations of the technology involved automation of routine tasks; only a small proportion of offices attempted to innovate.[28] Spreadsheets, the other major application of personal computers, may have led to the generation of more numbers and more "what if" analyses, but their effect on productivity has probably been incremental at best.

The companies that create information technology share the productivity and innovation problems experienced by the users of that technology. It is probably not coincidental that many of the firms that have pioneered process innovation—e.g., IBM, Xerox, AT&T, British Telecom, and Digital Equipment Corporation—are in the information technology industry. They have a dual purpose in using information technology to enable process innovation: (1) to derive benefit internally, and (2) to use the installation of computers and communication networks as a means to change customer processes. Those that choose not to do so may find their customers beginning to spend their capital on more proven enablers of change.

It is helpful in this connection to look to Japan, where the degree to which computers and communications are deployed may shed some light on the general role of IT in productivity. Although we know of no systematic effort to compare the prevalence of IT in Japan with that in the West, numerous informal accounts suggest a lower degree of IT use in Japan,[29] particularly in white-collar environments, where the *kanji* alphabet has hindered the implementation even of word processing for many years. Yet, by traditional measures of office productivity, Japanese firms do quite well. Japanese banks, for example, have much larger assets per employee than Western banks.[30] Some Western managers who have studied the information systems in their Japanese subsidiaries have suggested that the systems supply only basic functionality, rather than the functional "frills" they see in so many U.S. software applications,[31] which might account for a higher correlation between IT spending and productivity. Conclusive evidence about the IT–productivity relationship in Japan will have to await more rigorous study, but the information we do have does not diminish concerns about the role of IT in productivity enhancement.

Technology firms that concentrate more on hardware development than on the development of applications and new processes are a large part of the productivity/innovation problem. Andrew Rappaport and Shmuel Halevi, pointing out that computer hardware once lagged behind the identification of applications for enhancing work, but now far exceeds it, encourage the industry to "create and deliver new applications, pioneer and control new computing paradigms"—in other words, to change the way their customers do work.[32] Not to do so is very likely to invite sales to decline.

Firms that do develop new applications must do so in a new way. Organizations commonly tailor application packages to fit existing business practice, with the result that most business applications are functionally oriented; marketing systems solve marketing problems, sales systems sales problems, manufacturing systems manufacturing problems. Such "stovepiped" systems cannot support a process view of the organization; they imprison data within functions, so that new product designs cannot be released to engineering, sales data cannot be transferred to manufacturing, and customers for one product who might be customers for another product cannot be identified.

Even if business drivers were not making demands for process innovation, the state of information systems might be sufficient to drive companies in that direction. In the late 1980s, for example, many companies concluded that their systems needed to be cross-functional in order to meet the kinds of business and customer needs cited above. But few companies have been successful in integrating data and applications on an ad hoc basis. Had their model of the organization been process-based rather than functional, their systems would probably have been much more integrated from the beginning.

The combination of the need for a process view and the failure of most firms to identify measurable productivity or competitive benefits from IT investments[33] makes the use of IT for process innovation a virtual necessity. Yet, as an historical analysis of the relationship between IT and process thinking illustrates, the potential for IT-enabled innovation is only beginning to be realized.

THE IT–PROCESS–PRODUCTIVITY RELATIONSHIP

While we await sufficient research to enable us to understand why IT has failed to improve productivity at the level of the business unit, industry, or national economy, we might attend to the observations of a number of researchers who have suggested that the likely cause is a failure to take full advantage of IT's capacity to change the way work is done. "U.S. business has spent hundreds of billions of dollars on computers, but white-collar productivity is no higher than it was in the late 1960s," William Bowen writes in *Fortune*. He adds, "Getting results usually entails changing the way work is done, and that takes time."[34]

Process improvement and innovation are the best hope we have for getting greater value out of our vast information technology expenditures, yet neither researchers nor practitioners have rigorously focused on business process change as an intermediary between IT initiatives or investments and economic outcomes (see Figure 3-1). The implicit assumption that IT enables existing processes to be speeded up or performed with fewer resources has undoubtedly been correct in some environments, but the lack of visible macroeconomic benefit from IT suggests that the assumption needs to be made explicit and tested.

"It is not yet clear how interactive, computer-based tools affect basic missions and task processes in white-collar work groups, and how these in turn can affect organizational performance," one researcher writes in an overall review of IT's effect on white-collar work. He concludes, "Although examples of dramatic technology-related benefits have surfaced in the literature, we are still unable to define the relationship between the capabilities of flexible, powerful computer systems and the performance of organizations in the information economy."[35] He might have made the same comments about blue-collar processes. IT's relationship to manufacturing processes, and the effect of that relationship on manufacturing firm success, is also little understood.

To our knowledge, only one research study, based in the insurance industry, has treated process as a key factor in understanding the economic benefit of information technology.[36] That study attempted to determine the relationship between IT initiatives in the insurance distribution channel and increased revenues from those channels. Preliminary analysis found no relationship; insurance agents with technology sold no more insurance than those without it. Subsequent analysis, however, found that agencies

Figure 3-1　The IT–Process–Productivity Relationship

| Information Technology Initiatives | → | Process Change | → | Economic Outcome |

that changed their work processes to take better advantage of the technology significantly increased their insurance revenues.

Our objective here is not to use empirical research to confirm or deny the role of process change in IT's impact on economic performance, but rather to establish the overall importance of process thinking and help firms understand how process change can be accomplished. If these objectives are achieved, researchers can begin to include process innovation and improvement as key intermediate factors in studies of the benefits of IT investments.

In other words, both researchers trying to understand the benefits of IT and managers trying to maximize the value of IT must begin to think of process change as a mediating factor between the IT initiative and economic return (Figure 3-1). To do so can lead to a radical change in perspective. No longer, for example, would we expect an IT investment in itself to provide economic return. We would recognize that only change in a process can yield such benefits and that IT's role (together with other factors discussed below) is to make a new process design possible. Managers seeking returns on IT investments must strive to ensure that process changes are realized. If nothing changes about the way work is done and the role of IT is simply to automate an existing process, economic benefits are likely to be minimal.

Curley and Henderson, although they do not address process change specifically as a mediating factor, point to the need to manage "job and organizational impact" in order to achieve economic benefits.[37] Moreover, they argue that change must be managed on multiple levels of the organization, from the work of individuals to the processes of groups to the strategic initiatives of the enterprise. Combining this multilevel approach to benefit management with the process-mediated model described above yields the matrix depicted in Figure 3-2.

A specific application, as Curley and Henderson point out, can benefit all levels of the matrix. Taking a system for enabling a direct distribution channel in an agricultural chemicals firm, they demonstrated that laptops supported greater interaction between individual sales people and customers, eventually resulting in greater sales. Other aspects of the system held benefits for the work group and enterprise levels of the organization.

Managers designing and implementing new systems should strive to identify application benefits throughout the matrix. In

Figure 3-2   A Framework for Maximizing IT Impact

	IT Initiative	*Process Change*	*Economic Outcome*
Individual	Laptop System	Sales Call	Sales
Work Group	Product Database	Product Movement	Product Management
Business Unit	Product Management System	Channel Relationships	Competitive Position

several cases in which we have used the matrix with managers, we have found that they have been able to retroactively discuss changes in processes for which applications have already been created, and prospectively design information systems development initiatives to increase the likelihood of process innovation.[38]

If the relationship between IT and performance is affected by process change, it is likely that it is affected by the presence or absence of other enablers of process change. Few firms, for example, think to deploy IT in conjunction with another key change lever, new approaches to human resource management. A quantitative analysis of IT investment in the automobile industry, for example, found that the investments yielded no productivity or quality gains unless matched by similar levels of human resource innovation.[39] Research in this area, exploring, for example, the relationships between IT initiatives and the use of self-managing teams in organizations, is beginning to emerge.[40]

IDENTIFYING ENABLERS OF PROCESS INNOVATION

If IT's potential for business change is to be achieved, it must be viewed as an enabler of process innovation. As noted above, IT is only one of several enablers that typically work in concert to bring about change in processes. At some early point in a process innovation initiative, specific enablers must be identified. Activities through which these enablers are identified are listed in Figure 3-3.

The identification of change enablers must consider both what is possible and the constraints imposed by current technology and

Figure 3-3   Key Activities for Identifying Change Enablers

- Identify potential technological and human opportunities for process change
- Identify potentially constraining technological and human factors
- Research opportunities in terms of application to specific processes
- Determine which constraints will be accepted

organization. Change enablers must then be analyzed to determine the degree of freedom a company has in implementing new technologies and organizational forms given its current state. This is essential for cost/benefit analysis and migration planning. For example, a company might decide to institute a case manager position—an empowered individual at the customer interface aided by a powerful workstation—as part of a process innovation. But a study of the required skill set of employees reveals that retraining would not be possible; implementing a case manager model would require replacing the current staff, which is not an option under the company's policies. Similarly, proposed technology changes might entail substantial overhaul of the installed base of applications, which might be financially prohibitive. Finally, a company may be constrained by the need to continue to support existing products or services, for example, insurance policies issued under existing processes. Such product "legacies" may be difficult to support given a radically new process.

Having identified potential opportunities and constraints, their relevance to the process under analysis must be determined. Opportunities must be researched to determine how the technological or human innovation might be employed in the process. Constraints, on the other hand, are usually better examined through discussion than research. Which of the constraining factors should be accepted as constraints, and which should the organization try to overcome? Analysis at this point is still quick and high level, but it provides a better understanding of which enablers will become part of the process vision and how they will be employed in the process.

INFORMATION TECHNOLOGY AS AN ENABLER OF PROCESS INNOVATION

The conventional wisdom, even among advocates of process thinking, is that the process should be designed before investigating enabling technology or systems. For example, the approach of Westinghouse's Productivity and Quality Center, a leader in process thinking from the quality tradition, was always "first think about the process, then think about the system."[41] This is a natural reaction to the problem of automating poor processes, but it goes too far in the other direction.

Even those who advocate a strong relationship between business processes and information systems (e.g., proponents of information engineering[42]) typically focus on systems and technologies that help to implement, rather than enable, a process. The goal of information engineering is to describe an already-conceptualized process in informational (or, more accurately, data-oriented) terms so that a system can be rapidly and rigorously constructed to support the new process design. This is, as will be discussed more fully later, a valid objective, although it entails certain pitfalls that should be avoided.

But IT (and other enablers) can play an even more important role in process innovation (see Figure 3-4). When we understand how companies in many industries have used technology in innovative ways to improve their processes, we can better design new processes.

For example, any firm that intends to design a sales reporting

Figure 3-4   The Role of IT in Process Innovation

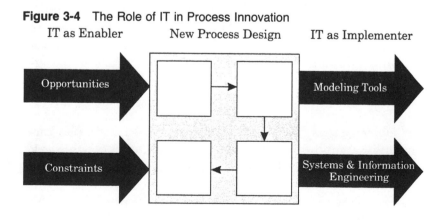

IT as Enabler   New Process Design   IT as Implementer

Opportunities   Modeling Tools

Constraints   Systems & Information Engineering

and analysis process should be aware of how Frito-Lay uses IT in its processes.[43] Any insurance company considering a new underwriting process should be cognizant of the many efforts to apply expert systems to underwriting decision-making processes. We know of no major process in which IT has not been used by some firm somewhere to achieve radical improvement.

To suggest that process designs be developed independently of IT or other enablers is to ignore valuable tools for shaping processes. A sculptor does not take a design very far before considering whether to work in bronze, wood, or stone. A process designer pursuing innovation should consider all the tools that can help to shape or enable the process, and IT and the information it provides are among the most powerful.

It is possible, of course, to take enablement of process innovation too far. IT and other enablers should never be employed for their own sake. To redesign a specific process in order to take advantage of imaging technology, as one company did, is an example. A process design should be enabled, not driven by, a particular change lever.

Enablement can have both positive and negative connotations; a process can be enabled or disabled by a particular tool. IT, for example, both provides opportunities for and imposes constraints on process design. Opportunities involve using technology in ways new to the company or industry to achieve process innovation. Constraints are those aspects of existing technology infrastructure that limit the possibilities for innovation and cannot, for whatever reason, be changed in the relevant time frame. In the next section, we analyze IT as an opportunity for innovation at both a general level and in terms of specific IT applications, and then consider how IT can shape processes as a constraint. We examine several examples in which IT acted as a key factor in enabling or disabling process innovation.

IT AND OPPORTUNITIES FOR PROCESS INNOVATION

Opportunities for supporting process innovation with IT fall into at least nine different categories (see Figure 3-5), which presume an overarching business objective of cost reduction, time elimination, and so forth.

The categories reflect the specific means by which these business objectives are achieved. Each is explained below.

Figure 3-5   The Impact of Information Technology on Process Innovation

Impact	*Explanation*
• Automational	Eliminating human labor from a process
• Informational	Capturing process information for purposes of understanding
• Sequential	Changing process sequence, or enabling parallelism
• Tracking	Closely monitoring process status and objects
• Analytical	Improving analysis of information and decision making
• Geographical	Coordinating processes across distances
• Integrative	Coordination between tasks and processes
• Intellectual	Capturing and distributing intellectual assets
• Disintermediating	Eliminating intermediaries from a process

Automational.   The most commonly recognized benefit of information technology is its ability to eliminate human labor and produce a more structured process. This opportunity, long understood in manufacturing, is the province of robotics, cell controllers, and so forth. In service environments, where processes are frequently defined by document flows, automational opportunities increasingly rely on imaging systems that remove paper from the process, frequently accompanied by "work flow" software that defines the paths images follow through a process.[44] In telephone-intensive processes such as telemarketing, order fulfillment, or customer service, automational control is typically provided by a telephone switch's automated call distributor (ACD).

Informational.   As Shoshana Zuboff has pointed out, information can be used not just to eliminate human labor from a process, but also to augment it.[45] Information technology can be used within a process to capture information about process performance, which can then be analyzed by humans (or, for that matter, by other information technologies—see IT's "analytical" dimension below). Zuboff's best-known example of the informating role of technology was in a paper mill, in which operators used computer-based tools to capture information on resource and energy consumption and then tried to optimize consumption levels based on the data and their expertise.

Sequential. IT can enable changes in the sequence of processes or transform a process from sequential to parallel in order to achieve process cycle-time reductions. This opportunity is at the core of the concurrent engineering phenomenon in product development. Given a well-structured design database and computer-aided engineering tools that can exchange design drawings, it is frequently possible to design in parallel components that once had to be designed sequentially. Kodak used this approach to reduce radically the design and development cycle for the single-use 35mm camera. In the insurance industry Phoenix Mutual Life employed a sophisticated computer system for work flow control to create from a totally sequential underwriting process a new process that is sequential in parts, parallel in parts, and can be reconfigured around bottlenecks such as vacationing employees. This new process, which is much more complex than its predecessor, allows the firm to issue an estimated 70% of its policies overnight.[46]

Tracking. To effectively execute some process designs, notably those employed by firms in the transportation and logistics industries, requires a high degree of monitoring and tracking. Federal Express, for example, scans a package as many as 10 times in order to be able to locate errant packages readily and to respond to customer queries regarding package status,[47] and satellite-based tracking systems enable a number of trucking firms to know the precise location of each truck in their fleets.[48] Opportunities for tracking also occur in nonlogistical processes. Johnson & Johnson's new centralized research management facility, for example, employs a database that enables executives to track the progress of drugs through the research and development cycle. Knowing the status throughout the pipeline permits the firm to avoid bottlenecks (e.g., too many drugs entering the clinical trials phase at once) and eliminate drugs that show less promise at particular points in the pipeline.

Analytical. In processes that involve analysis of information and decision making, IT can bring to bear an array of sophisticated analytical resources that permit more data to be incorporated in and analyzed during the decision-making process.

American Express's well-known expert system for authorizing credit card purchases, the Authorizer's Assistant, draws on more information and makes fewer bad credit decisions, and does so in less time, than most of the company's human authorizers. Risk analysis is particularly important to Progressive Insurance of Cleveland, Ohio, which uses IT systems to identify the lowest risks among the pool of high-risk drivers it insures in order to be able to offer drivers somewhat lower premiums than its less discriminating competitors. In a somewhat different vein, Rank Xerox U.K. has made its management processes much more informational and fact-based by using IT to digest operating and financial performance data, standardize its format, and distribute it to managers on demand. According to the firm's managing director, this process has enabled dramatic improvements in managers' understanding of the business and drastic reductions in time spent on routine meetings.[49]

Geographical.   A key benefit of IT dating to the invention of the telegraph has been the ability to overcome geography. Global companies are increasingly finding that their processes must execute seamlessly and consistently around the world. Both Ford and General Motors, for example, now use computer-aided tools to design, source, and manufacture components in different countries, and United Parcel Service and other international air freight companies use worldwide networks to send information to customs services so that shipments can be cleared before arriving. Digital Equipment Corporation's Enterprise Services consulting business was developed through planning sessions conducted over a worldwide electronic mail and computer conferencing system.

Integrative.   More and more, companies that are finding it difficult to radically improve process performance for highly segmented tasks split across many jobs are moving to a "case management" approach. In this type of process, an individual or team completes, or at least manages, all aspects of a product or service delivery process. We have seen this approach used in the telecommunications industry for circuit provisioning, in insurance for policy underwriting, in banking for commercial loans, and in hospitals for patient care management. In all these applications, information on various aspects of the process stored in databases spread throughout the organization is consolidated in a desktop workstation. We discuss this further in relation to customer-facing processes in a later chapter.

Intellectual.   Many annual reports cite employee knowledge and experience as a firm's greatest assets, but seldom are they well managed. Moreover, knowledge-intensive activities often are not treated as processes. Nevertheless, a number of companies are beginning to try to capture and distribute knowledge more broadly and consistently. Ford, for example, has implemented a global database of process knowledge for producing electronic components that can be easily accessed by any division interested in the experiences of other parts of the organization.[50] American Airlines is attempting to build a database of customer service practices and procedures that can be accessed by customer service representatives at any airport.[51] Finally, a number of the Big Six accounting firms have developed networks of information on tax and accounting issues. In each of these cases, the goal is to make expert knowledge available across an entire firm.

Disintermediating.   It is becoming increasingly clear in many industries that human intermediaries are inefficient for passing information between parties, particularly in relatively structured transactions such as stock brokerage, parts locating, and even finding a home. Consequently, many firms and stock exchanges are attempting to establish automated exchanges. One analysis of the New York Stock Exchange suggests that electronic trading could save stock buyers and sellers hundreds of millions of dollars annually.[52] Electronic brokerage databases have also been employed for such diverse products as aircraft parts, gems, used computers, and vacation homes.[53] Television is being used to sell a variety of products and services, even, in the Boston area, foreclosed real estate. A computer-based network of house images is being used to reduce the time prospective buyers spend seeing houses. In markets where there are high levels of choice and many participants, IT is invariably useful in connecting buyers and sellers and helping them exchange information about purchase transactions.

A process team thinking about how IT might be used to leverage a process should first identify which of the opportunity categories above are relevant, and then, if possible, find examples of how other firms, either within or outside the industry, have used IT to accomplish the same or a similar objective. The practice of visiting firms using IT in similar processes in order to understand process design and implementation issues—a form of benchmarking we call "innovation benchmarking" to distinguish it from its more measurement-oriented forms—also works with specific applications of IT, as described below.

GENERIC PROCESS APPLICATIONS OF IT

For process designers to incorporate IT capabilities in their process designs, the technology must be understood. Nevertheless, although technologists may be numbered among a process team's members, most designers should be individuals whose primary expertise is in the business. It is useless to speak in terms of raw IT capabilities when trying to link IT capabilities to process objectives.

For example, a "100 MIPS (millions of instructions per second) processor" or "200 gigabit optical fiber network" will mean little to the average process team member. Nor will comparisons to existing, more familiar technologies—e.g., "the mainframe on a desktop" or "the ability to carry 400,000 concurrent voice conversations." Such expressions do little to establish a connection between a process team's performance objectives and the tools that can enable them.

Rather, the capabilities of IT should be phrased in terms of application to common, or generic, business problems.[54] These generic applications are not a single technology, but a package of hardware, software, information, and communications designed to yield some useful functionality. For example, rather than discuss expert systems, a team should consider such generic applications as logistical planning, automated diagnosis, product choice, or decision analysis systems. Each of these applications may exploit expert systems technology, but may also take advantage of more powerful desktop workstations and recent advances in local area networking. Most important, these applications are means of solving business problems, not technologies looking for uses. Unless process innovation team members are themselves technologists, or do not trust the technologists' analysis, they need not be concerned with underlying technologies.

But technologists do have an important role to play in preparation for the process team's deliberations. For each major process under analysis, they should assemble a relevant set of generic applications. The set should err in the direction of inclusiveness; any application that might be relevant to the process should be included, and marginally relevant applications should be filtered out by the team rather than the technologists.

For each application, the technologists should provide a description, a listing of the underlying technologies that make it possible, an overview of the limits of its capabilities (both now and in the future; five years is the normal limit for accurate prediction), and some exemplary process problems that the application has helped to solve.

Although present capabilities are most important to a process team, it is helpful to make team members aware of medium-term future capabilities of an application. A number of firms have begun to design new processes well before the relevant technologies have fully matured. Frito-Lay, for example, planned to use handheld computers for sales reporting several years before such devices were capable of supporting the company's application.[55] Similarly, Citibank planned branch strategies and teller processes well before the automated teller machine was ready for customer use.[56]

To illustrate the relationship between process designs and generic applications, we will discuss potential generic applications and the power of IT to transform specific processes within firms. Generic applications are presented for three key processes—product development, order fulfillment, and logistics.

Generic Applications for Product Development Processes

The key challenges in product development processes—increasing the speed of design and prototyping, simulating process performance, tracking product status, coordinating product design information across components and products, and making decisions about resource allocation and market rollout—are addressed by at least five key applications of IT.

Automated design.   IT supports rapid design and prototyping of many components and processes, in both two and three dimensions, for both manufacturing and service enterprises. Physical models of designs can be produced in minutes or hours, and in limited design domains it is already possible to combine automated design with expert systems that evaluate alternative designs and select those that best meet specified criteria. Xerox, for example, has developed an expert-system design tool that optimizes paper path designs for new copiers. In the mid-term future, we expect to see design systems that will enable firms to optimize product and component manufacturability, and the beginnings of automated design, in which critical parameters and functionality objectives are input and a prototype design produced. We are also beginning to see the emergence of standards that would support the exchange of graphic design and product information across different hardware and software platforms.

Simulation systems.   Simulation technology enables product and process designers to simulate the execution of designs in increasingly realistic and complex settings. In product design, 3-D graphics workstations can display component movement in relation to other components and simulate stress loadings and effects of hostile environments; in process design, the implications of resource constraints and the ability of processes to run faster and to handle increased volume can be interactively simulated. Emerging "virtual reality" capabilities will provide for realistic, almost lifelike simulation of process environments. Information systems prototyping and screen-painting systems enable process designers to explore what it is like to use a system before building it.

Tracking systems.   In complex product-development processes, managers must be able to track continuously the status of particular products or projects in the development cycle. Project management systems are a variation on tracking systems, which can include such product development information as: project timelines; the names of individuals working on particular products (the purpose of a tracking system used by Cypress Semiconductor[57]); current and cumulative resources expended on a product and dependency relationships among these resources; problems associated with a product's development or eventual manufacture and use; and market feedback about a product's prospects. Future enhancements to tracking systems are expected to support graphic representation of product pipelines, provide early warnings of resource constraints, and facilitate ease of use and understanding by nontechnical managers.

Decision analysis systems.   Deciding when to apply additional resources, when to send a product to market, and when to cancel a development project are key business issues in product design processes.[58] A decision analysis system that calculates likely financial returns from product development investments at various stages of the product life cycle can inform such decisions. A prospectively oriented tool that employs expert systems to help with product introduction decisions was recently introduced,[59] but a retrospectively oriented system that includes results from past product-development resource-allocation decisions and the financial implications of those decisions would be an even more powerful tool.

Interorganization communication systems.   For many firms, a key product-development application is simply cross-organization communication. Product designers need to be able to exchange ideas for new products, day-to-day messages about progress, and actual product designs. At one level, intraorganizational communication involves common applications for electronic messaging and conferencing (e.g., bulletin boards[60]). At a less technological level, key issues in exchanging design information often revolve around information and data management—that is, establishing common component numbers and product information standards throughout an organization, building databases of product designs and information, and securing adherence to design and technology standards.

Though by no means the only generic applications relevant to product development processes, these five suggest the types of applications to be investigated and the roles they might play in process innovation. Because product development processes are sufficiently different across industries and companies, some tailoring of these and other relevant generic applications will undoubtedly be necessary.

Nor are generic applications of IT in product development exclusive to product design. IT can play a very important role in designing, simulating, making resource decisions, and communicating about process designs as well. We discuss this set of capabilities in a later chapter.

Generic Applications for Order Fulfillment Processes

The process of moving from a customer request or proposal to the collection of revenue is being examined by many companies today. Their goals are to speed delivery to the customer, increase customer satisfaction with the order, and eliminate costs and other resources consumed by the process. IT's role in the process derives from the need to coordinate order management across multiple parties and speed communication. At least six generic applications have potential for transforming order fulfillment.

Product choice systems.   Customers often have difficulty selecting products from complex or technical product lines. Companies with such product lines can use expert systems or databases to develop product choice systems that match customer needs to specific product solutions. Systems have been developed that facilitate product choice in computer system configurations,[61] air conditioning systems, agricultural pesticides, and automobiles.[62] Such systems can be used directly by customers or through the intermediary of a sales person. A future version of this application may well be an "automated sales rep" capable of responding without human assistance to many customer needs.

Microanalysis and forecasting.   Predicting customer demand is a perennial problem in order fulfillment. Some firms, rather than build products to customer specs, prefer to anticipate customer requirements. This has long been possible at the aggregate level, but now, with customer identity numbers proliferating in industries from airlines to supermarkets, and multiple data sources at the store and neighborhood level, firms are attempting to predict microlevel demand from analysis of past purchase behavior. Consumer products firms are predicting how much toothpaste will be sold in a given store under a specific promotion, and airlines are manipulating per-seat pricing using highly complex yield-management systems and data from previous responses to promotions on individual routes. The long-term direction for this application is a marketing strategy tailored to individuals, and the decline of mass-market advertising and promotions.[63]

Voice communications effectiveness.   Given that many order fulfillment activities involve the telephone, IT applications that enhance the effectiveness of voice communication represent significant opportunities for process innovation. At the simplest level, voice messaging capabilities can substantially improve communication among customers, sales management, and sales representatives (although this level of improvement is closer to process improvement than process innovation). In telemarketing-oriented environments, this application can provide customer identification (through automated number identification) at the beginning of a call, automatically display customer history on the account representative's screen, and even transfer calls from location to location. Florist Transworld Delivery (FTD) uses such an application to enable florists maintaining normal daytime hours to be available to customers around the clock. After-hours calls are received at a central FTD service center, where they are handled by service representatives who see displayed on their screens the name and location of, and specials and prices offered by, the florist being called. The local florist appears to be offering 24-hour service, but bears little of the attendant cost.[64]

Electronic markets.   Electronic markets, like other applications described here, can be simple or complex. Simple forms of this application include electronic catalogues accessed via customer terminals; more complex versions involve bidding, auction, and spot-pricing systems with automated clearing capabilities. A number of third-party firms (e.g., Compuserve and Prodigy) are attempting to establish videotext-based electronic markets for a variety of products and services.

Interorganizational communications.   Electronic data interchange (EDI), the well-known order management IT application, is a relatively primitive form of interorganizational communications, primitive because it usually involves only simple purchase and delivery transactions such as invoices or bills of lading. Used alone, EDI seldom leads to innovation. Transmitting such routine information, whether by mail or electronic networks, is only a small part of the order management process. Opportunities for innovation lie in combining EDI with changes in the processes that lead up to the generation of transactions. Technologies and standards being developed for EDI will enable companies to exchange in multiple forms (voice, data, video) information on product characteristics and usage as well as routine transaction data.

Textual composition. It is becoming increasingly common for computers to generate text with wording conditional on target readers. In the order management process, textual composition is most frequently applied to the generation of automated proposals. A number of firms in the IT industry have developed applications that enable their sales representatives to answer questions about customer situations in a structured dialogue and print out tailored proposals. This application will become increasingly common as firms attempt to make the previously unstructured process of proposal generation more efficient and less time consuming.

Companies are only beginning to tap the potential of IT for process innovation in order management. As they do so, they transform not only their own processes, but also those of their customers. The difficulties of innovating with processes that cross organizational borders are explored more fully when we discuss customer-facing processes in a later chapter.

Generic Applications for Logistical Processes

Logistical processes are concerned with the movement of goods within a firm or between a firm and its suppliers and customers. Order management, inventory and materials management, and service delivery in transportation firms are among the specific processes that involve logistics. Among the generic applications of relevance to logistical processes are locational, recognition, asset management, logistical planning, and telemetry systems.

Locational systems. Key to logistical processes is knowing the location of materials or vehicles within a geographical network. Sophisticated computer and communications technologies are increasingly being used to rapidly and accurately determine the location of business entities—people as well as goods—and are frequently combined with applications that provide alternative routings of vehicles or goods. Schneider National, the largest U.S. long-distance trucking company, for example, constantly monitors the location of each of its 6,700 trucks via satellite,[65] enabling route planners to deploy trucks quickly for unscheduled customer pickups. American President Lines employs a similar system for its ships, which enables it to monitor and relay to customers the exact arrival times of their shipments.[66] Cellular radio and pager technology-based locational systems that monitor the locations of key individuals are also emerging.

Recognition systems.   An adjunct to the need to locate objects is the ability to identify them accurately and quickly. This capability ranges from reading bar codes on objects, which has become pervasive in the consumer-goods retailing industry, to identifying objects through computerized pattern recognition. The latter capability, now used primarily in manufacturing, is being explored in transportation for shipment confirmation. Other technologies, such as biometric systems, can be used to recognize individuals for purposes of confirming their identity as a customer or employee.

Asset management systems.   To optimize the use of key assets in processes—be they physical goods inventory, human resources, or financial assets—companies must be constantly aware of the location, availability, and best use of those assets. "Smart warehouses" of physical goods track the movement of inventory in and out of company facilities; asset management, in the form of productivity measurement systems, measures and reports by computer the time it takes an employee to complete a task;[67] and automated cash management applications help companies make optimal use of liquid financial assets.

Logistical planning systems.   Although logistical planning systems for solving complex problems of routing, scheduling, and resource assignment have been available for a number of years, they have not been well suited to solving most logistical process problems. Their complexity and inaccessibility to nonspecialists greatly limited their application. Logistical planning will continue to be complex, but rule-based technologies such as expert systems can solve the simpler logistical problems of processes and construct and interpret more complex linear-programming models when necessary. Process innovation in logistical planning will therefore involve making planning tools accessible to process participants at their desks or workstations.

Telemetry.   Telemetry, the ability to monitor a process from a distance, is manifest in such wireless technologies as microwave and radio, which make it possible to measure and record information from instruments that are physically remote. Telemetry has engendered innovation in the reading of household utility meters by readers that record current levels of utility use broadcast by low-power radio devices. As such devices become smaller and cheaper, they could be deployed to any process that involves a need to record information rapidly about nearby objects (e.g., cars passing a toll booth or trucks passing a highway weighing station).

The generic applications listed for these three processes are representative. A firm that intends to use IT to enable process change must explore existing applications much more deeply. The concept of generic applications is an understandable way to focus a process team on how IT can enable innovation. A company that wants to implement a specific generic application in its processes is well advised to observe its successful functioning in another firm. This is another approach to innovation benchmarking. It usually makes no difference what industry or type of firm is observed; the point is to learn how IT was applied to, and implemented within, the process.

IT AS A PROCESS CONSTRAINT

Just as information technology can provide exciting opportunities for process innovation, it can also impose considerable constraints on process designs. It is easy to suggest that firms ignore existing systems and technology infrastructures in designing a new process, but it is seldom realistic to do so. Existing systems are often too expensive, complex, and embedded in an organization to simply assume them away. Instead of pretending to have a clean slate, firms should acknowledge the constraints existing systems impose on a new process, understand their implications, and make the best of them.

Strategic planners at a regional Bell operating company, for example, were considering a radical change in key business processes. The planners wanted to ignore industry conventions in designing processes that they were convinced would not even resemble those of other telephone companies. When asked about systems support for these processes, the planners replied, "We assume we will continue to get all our systems from Bellcore" (the research and development organization for the regional Bell companies, which develops and maintains many common software applications). These managers did not perceive the conflict between their goal of creating radically new processes and the constraints imposed by existing, industrywide information systems.

Even if they had been aware of the conflict, it is unlikely that their firm would have abandoned the installed Bellcore base—worth tens or perhaps hundreds of millions of dollars—to develop entirely new systems to support its processes. Instead of ignoring the conflict, the firm should consider the process design implications of retaining the Bellcore systems. It is possible that substantial process improvement could be made without abandoning the old systems.

In a natural gas transmission company, process constraint took the form of a recently acquired inventory management application package. Although the company was redesigning its process for inventory management, its chief information officer did not want to adopt a clean slate approach to inventory management. "I signed a contract for that package the day I took this job a year ago," he said. "We hadn't decided to redesign inventory management then. The package cost several million dollars and we can't just throw it away. I wouldn't make the same purchase today, but we can't act as if we have total freedom from the systems standpoint." At the time the CIO was interviewed, he was working with the process team to identify the level of freedom the new system would allow in redesigning the process.

At a money-center bank that has invested heavily in IT-enabled process innovation, the information systems organization begins to investigate packages that might meet process support needs well before the process design is complete. "We used to buy packages and modify them to suit our idiosyncratic processes," noted the director of IT planning. "Now we are just as likely to modify our process to fit the package."

Because the processes assumed by a package may be more logical and rational than those employed by many companies (because they are designed to be used in many different environments), this process design "constraint" may be a liberating one. Yet some packages do not even accommodate process-oriented thinking. Since most businesses are functionally organized, many packages are designed to support only specific functions.

A package or existing system that may be a given or constraint in a new process design should be evaluated to determine which process elements are implicit within or assumed by the system. The following are among the aspects of the system that should be analyzed:

- Who are the intended users of the system?

- What are its inputs and outputs?

- What process tasks is the system designed to support?

- How difficult is it to add task functionality to the system?

- What interfaces to other systems are possible?

- What processes do other firms use with the system?

When a process extends across organizational boundaries into customer and supplier organizations, it may be impossible to assume a clean slate of systems support. One cannot expect a customer to change systems to better supply one's firm with process information. Instead, the external systems should be analyzed for process degrees of freedom, just as an internal constraint system would be.

It is also possible to evaluate existing or proposed systems solutions for processes on the basis of the process opportunity categories described above. A firm that discovers, for example, that it cannot overcome geography and extend a process across Europe because each European division has its own incompatible systems may deem this too great a constraint to live with. If a process cannot be made more analytical because the architecture of existing systems cannot accommodate expert technologies, perhaps the process must be done outside the existing architecture.

Considering the existing systems environment as a process constraint may seem to limit the prospects for radical innovation, and indeed, if an organization chooses not to change many of its systems, the possibilities for process innovation are restricted. But rationally analyzing system constraints at least makes these trade-offs conscious. Rather than assuming a clean systems slate at the beginning of a process and then later getting bogged down in existing systems, the analysis of constraints tailors the process to a systems environment from the beginning.

SUMMARY

By now it should be obvious that information technology can have important implications for key business processes. Our goals in this chapter were the following: (1) to discuss the disappointing history of IT in enabling process change, and the threat to productivity when IT does not lead to such change; (2) to establish that IT should explicitly be considered as a change lever or enabler of process innovation before selecting a specific design; (3) to point out that IT can leverage process innovation in multiple ways; (4) to persuade even ardent technologists that the generic application is a useful way of connecting specific IT capabilities and process objectives; and (5) to convince process teams that IT constraints should also be explicitly factored into process designs.

IT capabilities—when described and discussed in nontechnical terms, as they were here, and applied to process problems—can work miracles by the standards of previous generations. How else but through this technology can we manage our processes globally, instantly, efficiently, and correctly? It is clear that no other tools are comparable. But technologies alone cannot work miracles. Innovations in the use of computers and communications must be combined with innovations in how information is used and structured. The next chapter discusses this relationship and examines the use of innovative information management practices as an enabler of innovative process designs.

Notes

[1]JoAnne Yates, *Control through Communication: The Rise of System in American Management* (Baltimore, Md.: The Johns Hopkins University Press, 1989).

[2]Leonard F. Vogt, "International Shoe Company," in George Schultz and Thomas L. Whisler, *Management Organization and the Computer* (Glencoe, Ill.: Free Press, 1960).

[3]Sanford L. Optner, *Systems Analysis for Business Management* (Englewood Cliffs, N. J.: Prentice-Hall, 1960): 11.

[4]See, for example, advertisements for Honeywell and New England Mutual Life in the *Boston Globe* (April 17, 1969): 24.

[5]John Graham, *Systems Analysis in Business* (London: George Allen & Unwin, 1972).

[6]John E. Hines, "A Division of General Electric Company," in Schultz and Whisler, *Management Organization and the Computer*, 153–155.

[7]Duncan Copeland and James McKinney, "Airline Reservations Systems: Lessons from History," *MIS Quarterly* 12:3 (September 1988): 353–372.

[8]Benn Konsynski and Michael Vitale, "Baxter HealthcareCorporation: ASAP Express," 9-188-080. Boston: Harvard Business School, 1988, rev. 1991.

[9]Eric K. Clemons and Michael C. Row, "McKesson Drug Company: A Case of Study of Economost—A Strategic Information System," *Journal of Management Information Systems* (Fall 1988): 141–149.

[10]E. J. Kane, "IBM's Total Quality Improvement System," IBM Corporation, unpublished manuscript, p. 5.

[11]H. James Harrington, *Business Process Improvement: The Breakthrough Strategy for Total Quality, Productivity, and Competitiveness* (New York: McGraw-Hill, 1991). See also his book *The Improvement Process: How America's Leading Companies Improve Quality* (New York: McGraw-Hill, 1987).

[12]See, for example, George D. Robson, *Continuous Process Improvement* (New York: Free Press, 1991).

[13]John Musgrave, "Fixed Reproducible Tangible Wealth in the United States, 1982–85," *Survey of Current Business* (Washington, D.C.: U. S. Department of Commerce, Bureau of Economic Analysis, August 1986).

[14]Steven S. Roach, "Economic Perspectives," *Morgan Stanley* (July 14, 1988).

[15]Robert Kominski, "Computer Use in the United States: 1989," Series P-23, No. 171 (Washington, D.C.: U. S. Department of Commerce, Bureau of the Census, Current Population Reports, March 1991).

[16]Lester C. Thurow, "Foreword," in Michael S. Scott Morton, ed., *The Corporation of the 1990's: Information Technology and Organizational Transformation* (New York: Oxford University Press, 1991): v–vii.

[17]Steven S. Roach, "Economic Perspectives," *Morgan Stanley* (January 1991): 6–19.

[18]Roach, "Economic Perspectives," July 14, 1988.

[19]Gary W. Loveman, "An Assessment of the Productivity Impact of Information Technologies," working paper 90s: 88-054, MIT Sloan School of Management, Management in the 1990s, July 1988.

[20]M. N. Baily and A. Chakrabarti, *Innovation and the Productivity Crisis* (Washington, D.C.: Brookings Institution, 1988).

[21]Suzanne Iacono and Rob Kling, "Changing Office Technologies and Transformations of Clerical Jobs: A Historical Perspective," in Robert E. Kraut, ed., *Technology and the Transformation of White-Collar Work* (Hillsdale, N. J.: Lawrence Erlbaum Associates, 1987): 53–75.

[22]Roslyn L. Feldberg and Evelyn Nakano Glenn, "Technology and the Transformation of Clerical Work," in Kraut, ed., *Technology and the Transformation of White-Collar Work*, 77–97.

[23]Raymond R. Panko, "Is Office Productivity Stagnant?" *MIS Quarterly* (June 1991): 191–203.

[24]Thomas D. Steiner and Diogo B. Teixeira, *Technology in Banking: Creating Value and Destroying Profits* (Homewood, Ill.: Business One Irwin, 1990).

[25]Paul Strassman, *The Business Value of Computers* (New Canaan, Conn.: The Information Economics Press, 1990).

[26]Kathleen Foley Curley and John C. Henderson, "Valuing and Managing Investments in Information Technology: A Review of Key Models with a Field-Based Framework for Future Research," ACM/OIS Conference: Value Impact and Benefits of Information Technology, Minneapolis, Minn., May 1989.

[27]Kominski, "Computer Use in the United States: 1989."

[28]James C. Taylor, "Job Design and Quality of Working Life," in Kraut, ed., *Technology and the Transformation of White-Collar Work*, 211–235.

[29]See, for example, Tatsumi Shimada, "The Impact of Information Technology on Organizations in Japanese Companies," in *Management Impacts of Information Technology: Perspectives on Organizational Change and Growth*, Edward Szewczak, Coral Snodgrass, and Mehdi Khosrowpour, eds. (Harrisburg, Pa.: Idea Group Publishing, 1991), which describes lower rates of IT spending relative to revenues than in U.S. companies, and lower workstation use per capita.

[30]See "The Top 500 Banks in the World," *American Banker*, July 27, 1990, 18a–31a. Because of differences in the businesses and asset mixes of Western and Japanese banks, these figures should be interpreted with caution.

[31]For example, personal conversations with S. I. Gilman and Larry Ford, senior information systems executives at Ford Motor and IBM, respectively.

[32]Andrew S. Rappaport and Shmuel Halevi, "The Computerless Computer Company," *Harvard Business Review* (July–August 1991): 69–80.

[33]Curley and Henderson, "Valuing and Managing Investments in Information Technology."

[34]William Bowen, "The Puny Payoff from Office Computers," *Fortune* (May 26, 1986): 20.

[35]Tora Bikson, "Understanding the Implementation of Office Technology" in Kraut, ed., *Technology and the Transformation of White-Collar Work*, 155–176.

[36]N. Venkatraman and Akbar Zaheer, "Electronic Integration and Strategic Advantage: A Quasi-Experimental Study in the Insurance Industry," working paper 90s:89-072, MIT Sloan School of Management, Management in the 1990s, April 1989.

[37]Curley and Henderson, "Valuing and Managing Investments in Information Technology." In personal conversations, Curley and Henderson have agreed that the process change variable is a good substitute for job and organizational impact.

[38]This matrix, and the overall topic of process as an intermediating factor in returns from IT, are derived from a research project at Ernst & Young's Center for Information Technology and Strategy[SM] entitled "Realizing the Benefits of Process-Focused IT." Participants in the research included Kathleen Curley, Thomas Davenport, John Henderson, and Suzanne Pitney.

[39]Paul Osterman, "Impact of IT on Jobs and Skills," in Michael S. Scott Morton, ed., *The Corporation of the 1990's*, 220–243.

[40]For a case study, see Lynda Applegate, "General Electric Canada: Designing a New Organization," 9-189-138. Boston: Harvard Business School, 1989. For a more analytical approach, see Laku Chidambaram, Robert P. Bostrom, and Bayard E. Wynne, "The Impact of GDSS on Group Development," *Journal of Management Information Systems* 7:3 (Winter 1990–91): 7–26.

[41]Personal interviews in 1989 with managers of the Westinghouse Center for Productivity and Quality in Pittsburgh, Pa.

[42]See, for example, James Martin and Clive Finkelstein, *Information Engineering* vols. 1 and 2 (Carrforth, Lancashire: Savant Institute, 1981).

[43]Melissa Mead and Jane Linder, "Frito-Lay, Inc.: A Strategic Transition (A)," 9-1187-012. Boston: Harvard Business School, 1987; and Nicole Wishart and Lynda Applegate, "Frito-Lay, Inc.: HHC Project Follow-Up," 9-190-191. Boston: Harvard Business School, 1990.

[44]"Workflow: Automating the Business Environment," BIS CAP International (Norwell, Mass.: October 1990).

[45]Shoshana Zuboff, *In the Age of the Smart Machine: The Future of Work and Power* (New York: Basic Books, 1988).

[46]B. Birchard, "Remaking White Collar Work," *Enterprise* (Digital Equipment Corporation, Winter 1990): 19–24.

[47]Personal interviews with various managers in Federal's Strategic Integrated Systems group, September 1990.

[48]"Listen Up, 18-Wheeler, We Know You're in Tacoma," *New York Times* (March 10, 1991) Sunday, late edition—final, sec. 3, p. 9, col. 1.

[49]Personal interview with Vernon Zelmer, Managing Director, at Rank Xerox U.K. headquarters, May 17, 1991.

[50]Thomas A. Stewart, "Brainpower: How Intellectual Capital Is Becoming America's Most Valuable Asset," *Fortune* (June 3, 1991): 46–60.

[51]This system, called InterAAct, is partially described in Max Hopper, "Rattling SABRE—New Ways to Compete on Information," *Harvard Business Review* (May–June 1990): 118–125. Further detail is provided in Joyce Wrenn, "InterAAct—Creation of a Technology Platform," in *Information Empowerment*, Critical Technology Report C-4-1 (Carrollton, Tex.: Chantico Publishing, 1991).

[52]Jason Forsythe, "The Big Board: Boxed in by Automation," *InformationWeek* (May 20, 1991): 46–57.

[53]Benn Konsynski and Art Warbelow, "American Gem Market Systems," 9-189-088. Boston: Harvard Business School, 1988; Benn Konsynski and Art Warbelow, "Inventory Locator Service," 9-191-008. Boston: Harvard Business School, 1990.

[54]The concept of generic applications was first developed by the author in the context of a consulting study for Shell Oil Company. I am grateful to Shell for funding this work, which has proven useful in a variety of other settings.

[55]Mead and Linder, "Frito-Lay, Inc.: A Strategic Transition (A)."

[56]The task of technology futures planning was eased at Citibank by the fact that a bank subsidiary, Transaction Technologies, was developing the ATM.

[57]Brian Dumaine, "The Bureaucracy Busters," *Fortune* (June 17, 1991): 36–50.

[58]Charles H. House and Raymond L. Price, "The Return Map: Tracking Product Teams," *Harvard Business Review* (January–February 1991): 92–100.

[59]Peter H. Lewis, "Software to Help Introduce Products," *New York Times* (October 6, 1991), sec. 3, p. 8.

[60]See, for example, Lynda M. Applegate and H. Smith, "IBM Computer Conferencing," 9-188-039. Boston: Harvard Business School, 1990.

[61]Dorothy Leonard-Barton, "The Case for Integrative Innovation: An Expert System at Digital," *Sloan Management Review* (Fall 1987): 7–19.

[62]Michael Vitale, "General Motors Corp.: The Buick EPIC Project," 9-188-058. Boston: Harvard Business School, 1988.

[63]For more information on this phenomenon, see B. Joseph Pine II, *Mass Customization* (Boston: Harvard Business School Press, 1992).

[64]Elisabeth Horwitt, "May I Say Who's Calling?," *Computerworld* (May 27, 1991): 1–14.

[65]Agis Salpukas, "Computers Give Truckers an Edge," *New York Times* (May 25, 1991): 35.

[66]Jerome D. Colonna with Stephen Sabatini, "Keeping American President Afloat," *InformationWeek* (August 15, 1988): 28–32.

[67]This computerized monitoring may be oppressive to employees, however, and any productivity gains may be neutralized by resentment. For an example of overmanagement of human assets, see Nitin Nohria and J. Chalykoff, "Internal Revenue Service: Automated Collection System," 9-490-042. Boston: Harvard Business School, 1990.

Chapter 4
Processes and Information

The previous chapter examined information technology's role in making process innovation possible. But we have yet to consider the relationship between processes and information: how information supports, is produced by, and should be managed in business processes. Information—the words, numbers, images, and voices that impart meaning and inform the information consumer—is the focus of this chapter. We choose not to distinguish between information and such related terms as data, knowledge, wisdom, and so forth, but rather to view these alternate terms as information with differing degrees of interpretive value added. The point at which data has enough value added to become information is a matter for philosophy (if anything), not business.

It is difficult to fully disentangle information from information technology and systems. Like the first three chapters of this book, most analyses of the "information revolution" in business have focused on information technology, only occasionally separating the thing being manipulated (information) from the thing doing the manipulating (information technology). One reason for separating the two entities is that much information in organizations and processes—more than 85%, by some estimates—is not manipulated by information technology. But although it may be too unstructured to be captured or distributed by a computer, such information can still be a useful process input or output. Finally, when we don't separate information and information technology, information always seems to get short shrift in the analysis. This chapter is an attempt to restore some of the balance.

Because of the semantic confusion that surrounds information, a compromise term, "information management," meaning the overall management of a firm's entire information environment, has gained currency.[1] The term imparts equal weight to the patterns and valuation of information usage and to the systems and technology that enable usage. It is the most relevant term for a firm's information practices in relation to process innovation. Information is the most important of the information management components, and it plays an equally important role in process design and innovation. The information management orientation also possesses a base of theoretical and methodological knowledge that practitioners and academics have built up over time. In fact, the disciplines of library, information, document, and computer and decision sciences, and other, more business-oriented fields all have made contributions to the theoretical study of information.[2]

But despite much talk and writing about the "Information Age," few organizations have treated information management as a domain worthy of serious improvement efforts. Progress in managing information is rarely described or measured; providers of less-structured information in organizations (e.g., libraries, competitive analysis functions, and market research groups) often have relatively low status and confused reporting relationships; vast amounts of information enter and leave organizations without anyone's being fully aware of their impact, value, or cost. Information management is thus a natural target for a process orientation, and many executives we speak with feel implicitly that it will be key to their competitive success in the future. We believe that managing information in terms of its role in business processes is prerequisite to achieving that success.

The management of information being a speculative topic, this is a speculative chapter. Our objective is modest—to convince process-oriented managers and professionals of the value of thinking about information as a process entity. Much work awaits the academics, consultants, and managers who would further our understanding of the worth and management of this resource.

Here, we discuss information and processes in terms of four roles:

- as a supporting tool for business process innovation;
- as a focus of operational processes;
- as a focus of management processes; and
- in terms of approaches to its management within processes.

Information is used and thought about in different ways in each of these roles. In studying how information interacts with processes, whether embedded in and manipulated by information technology or not, one can begin to perceive future uses and directions in both of these domains.

THE ROLES OF INFORMATION IN PROCESSES

Information can play a number of supporting roles in efforts to make processes more efficient and effective. Just the addition of information to a process can sometimes lead to radical performance improvements. It can be used to measure and monitor process performance, integrate activities within and across processes, customize processes for particular customers, and facilitate longer-term planning and process optimization.

Process Performance Monitoring

The role of information in monitoring process performance has been familiar to industrial engineers and systems analysts since Norbert Wiener, who advocated information feedback loops in work systems. Information is also important to quality experts, who counsel that quality cannot be improved without knowing the quality of existing activities. Juran, for example, who speaks frequently about the importance of "sensors" in monitoring quality, observes that "operating forces" need sensing to keep them "in a state of conformance to goals."[3]

Garvin, in analyzing product-quality information management within the room air-conditioner industry, found that

> quality departments at the best and better plants were not distinguished by such traditional variables as large size, a large number of inspectors, a particular organizational form, the power to issue hold orders or stop the assembly line, or the range of performance tests. Rather, they were notable for their effectiveness in managing and monitoring information about quality.[4]

Garvin also discovered that Japanese air-conditioner manufacturers were generally much better at "generating, collecting, and reporting" quality data than their U.S. counterparts.[5]

The use of information in process monitoring is even more important when information technology is used to automate some aspect of the process. Computers are capable of capturing and reporting such information as the resources consumed by, and duration, output characteristics, and even cumulative cost, of processes. As Shoshana Zuboff explains, "The devices that automate by translating information into action also register data about those automated activities, thus generating new streams of information."[6]

The Japanese automobile industry offers an interesting example of viewing information as an enabler of process monitoring. Cars are manufactured in Japan under a system termed "lean production" by some researchers. Pioneered by Toyota, this powerful effort to enhance teamwork and efficiency generates information of a unique kind. "All information—daily production targets, cars produced so far that day, equipment breakdowns, overtime requirements, are displayed on . . . lighted electronic displays that are visible from every work station."[7] Summations of the interactive daily record are provided to senior managers to enable them to more clearly understand process quality and productivity throughout the plant.

General Electric's Salisbury, Maryland, electrical components manufacturing plant, renowned for its excellent manufacturing and order management processes, uses similar real-time displays to apprise workers and managers of performance figures. Because worker compensation is based on meeting short-term performance goals, the displayed information has immediate relevance.[8]

The benefits of such performance reporting systems are clearly tied to the real-time, fully accurate nature of the information they display. A recent study of managerial information use found that much computerized information is not used by managers because it is obsolete.[9] Managers preferred more immediate feedback, even if only an estimate, over accurate information delivered too late to act upon.

But real-time display of accurate and complete process information is appropriate only to processes and objectives that are adequately described by the simple counting of outputs, transactions, or defects. Other processes and objectives require more innovative and unobtrusive methods, such as sampling and tracking cases through processes, surveying process customers, and building information systems for the specific purpose of process monitoring. Process information can also be used over longer periods of time to analyze patterns of performance and optimize the design and execution of a process.

Information-Based Process Integration

Whatever the organization of a company and structure of its processes, there is a need to coordinate process activities across geography and time and pass information from one process to another. It is frequently said that information is the "glue" that holds an organizational structure together.[10] Information can be used to better integrate process activities both within a process and across multiple processes. Often, information gathered for one process proves to be useful in another. The role of information in process integration is perhaps best illustrated by examples. We offer several below.

We turn again to the Japanese automobile industry to illustrate how information originally collected in one process came to be valued and used in other parts of the business. Cars in Japan are mostly sold by door-to-door salesmen who develop strong relationships with their customers. The salesmen, who usually work for distributors in which the car manufacturers have an equity position, collect all information relevant to the ordering of a new car, including any comments on the performance of previous models and extensive data on the customer. This information is aggregated, analyzed, and distributed to designers, quality managers, marketers, and senior management.

Inbound consumer "800" numbers provide another example of information used as a process integrator.[11] These lines, originally set up to enable consumers to discuss product and service problems and questions directly with manufacturers, were usually established during "getting close to the customer" programs and are themselves interesting examples of technology-based service processes. But aggregated and analyzed, the mass of information that pours into a company over these lines constitutes both the "voice of the market" and a vast store of useful customer data.[12] General Electric's 24-hour answer lines, for example, annually receive more than three million calls. This remarkably extensive resource is analyzed and distributed across many of the firm's consumer-oriented divisions. It is also analyzed in the company's "Work-Out" process-improvement program, described elsewhere in this book.

Information needed to integrate existing processes can be either generated internally or purchased from outside providers such as commercial information firms or industry associations.

When Becton Dickinson's senior managers decided they wanted a much clearer vision of the company's relationships with its major hospital customers, they established a hospital information database. Data that paint a comprehensive picture of each of these clients are bought from external providers and merged with internal sales, marketing, and product data to create an integrated file for each hospital that can be accessed by multiple processes.

Not only information, but also information-oriented employees can serve an integrating role within and across processes. The manager of the Information Resource Center at a regional Bell holding company headquarters, for example, commissioned a survey to determine which services provided by the center were valued most by users. The responses were a great surprise to the center's staff. The most valued service was the staff's awareness of who was working on what project; staff members often were involved in the research for these projects and would refer internal clients to others working on projects with similar features. The survey respondents viewed the group as a "human PBX system."

Process Customization

A key role of information from the process customer perspective is to enable the tailoring of process output to customer needs, a role termed "mass customization" by the IT and business presses.[13] With vast stores of customer information and powerful technologies to search through and manipulate it, there is no longer any excuse for serving mass markets with unvarying products. There is no longer any reason a firm, whether it sells toothpaste, airline seats, magazines, or consulting services, cannot know enough about its customers to be able to tailor its product (or service) offerings to individuals. Firms need to recognize that there is no longer a market, only individual customers.

There are many examples of the successful use of information in process customization. Frito-Lay, for example, can create spicier chips for stores in particular areas of a city; almost any major airline can offer a frequent flyer better treatment than other customers; any telemarketing firm can now answer the telephone with the customer's name and immediately display a record of the customer's previous transactions with the firm; magazines can be bound selectively, with advertisements and editorial content designed to appeal to particular readers. The degree to which a process can deliver customized outputs has become an important indicator of its quality.

Any firm that can master the complexity of information can master the complexity of multiple customers and products. What are the secrets to making process customization work? Firms that have succeeded have mastered the basics of information management. They are able to categorize, store, retrieve, and maintain customer records with relative ease. These firms determined early what information they needed to be able to offer tailored products or services and gathered that information. Because they recognized the value of process customization, they recognized the importance of their information assets and were willing to invest in them. The Henry Ford era, in which processes created standard outputs in which the only customer choice was to buy or not buy the product, is over, no matter what the business.

Information-Oriented Processes

The output of a process can consist either of tangible, physical goods and services or information products, or a combination thereof. There are also two types of information-oriented processes: those designed to aid management decisions and activities; and those with operational objectives. Management processes, almost all of which have a strong information component, are discussed both here, with respect to information issues, and in a later chapter as the subject of process innovation in general. Even among information-oriented operational processes, there are both unstructured and transaction-driven types. The various types of information-oriented processes to be explored in this section are displayed in Figure 4-1.

Information-Oriented Management Processes

In general, the state of information-oriented management processes is not good. Little about the quality, topical focus, or distribution of information to management in most firms is worthy of emulation. Most firms concentrate on financial information generated from accounting systems, which, as has been widely recognized, is frequently misleading or useless for management purposes.[14] Although managers clearly need externally generated

Figure 4-1   Information-Oriented Process Types

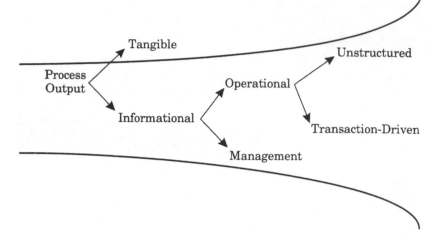

information about customers, competitors, and markets, most information processes, such as they are, continue to focus on internal information. Electronic distribution systems for management
information, such as the executive information system (EIS), are
often but slavish imitations of their paper-based predecessors.

A perhaps more philosophical answer as to why there are so
few successful information-focused management processes takes
account of the relationship between structured information and
management needs. As Henry Mintzberg, among others, has
pointed out: "Managers collect everything from impressions,
moods, and gossip to very hard data. They get information through
all kinds of channels and all kinds of ways, some of which happen
to lend themselves to machine processing, and a lot which do
not."[15]

Mintzberg's observation reflects one important reason for the
relative lack of impact of computers on senior management: the
sources of information needed by senior management—books, reports, and conversations, for example—are often not amenable to
computerization because of their unstructured and externally
generated nature. A recent study of managerial information use
in twelve manufacturing companies found that:

> The primary means by which information travels in the firms
> we visited is by interpersonal contact of some form, whether in

formal settings such as meetings or, as is more frequently the case, through informal interactions in the hall or the plant. Information exchanged in this manner is not limited in its characteristics to the expected gossip, intuition, or qualitative items. Indeed, one of the observations that struck us as unexpected was the fact that most numerical data appear to be passed by word of mouth first, with formal reports serving as later corroboration or reminders of what was transmitted orally.[16]

It is difficult even to construct a meaningful categorization scheme that would permit easy retrieval and perusal of this type of information. Most computer-based executive information systems fail to respond to the law of "requisite variety"—they do not draw their data from sources as varied as the complexity of a CEO's decision-making needs. In fact there has been little progress, either with or without computer systems, in executive scanning and acquisition of external information for decisions since Frank Aguillar wrote the first serious book on the subject more than 25 years ago. As he stated then, "Top management's understanding of the scanning process was found to be generally inadequate and there was little evidence of any coordinated, overall consideration of the problems involved."[17] Almost all the executive information systems implemented in firms today use one or two basic on-line sources for their external information—often a stock price service and news bureau—not because of technological limitations, but because of a failure to understand the information requirements of managers.

Because many executive systems tend to use information in a functional, rather than process-oriented, manner, few senior managers have information about how long it takes to develop a new product or the average time it took to fill customer orders during the previous month. The functional data such systems do provide are difficult to combine in a way that facilitates the monitoring of process performance. Until there are systems that support entire processes, process performance information must continue to be generated manually, by sampling, time reporting, walking a document through the process, and so forth. Only when they move toward a process orientation will companies begin to generate information that will support real-time analysis of how good their processes are.

Few organizations undertake specific initiatives to improve

their processes for generating management information. One exception is IBM Latin America, where the board of directors determined that it was not getting the right information for the types of decisions it was being called upon to make.[18] Getting current and historical market and economic conditions for Latin America's 24 countries is no small issue, given the differing levels of sophistication of those countries' own data-gathering mechanisms. IBM designated a task force, with representation from different internal functional groups, to review the situation. Instead of drawing upon existing information from current information-generating processes, the group designed new processes for generating the required information using state-of-the-art computer technology and extensive internal and external information on both financial and nonfinancial measures. The resulting on-line executive information system, LA Book, went much further than expected by the board. But to achieve what it did, IBM Latin America had to create, both within the Latin American countries and at its own headquarters, new positions whose functions were to acquire, capture, and analyze the information required by the system. These analysts extract data from individual corporate functions and combine them into a process-oriented view. This system has proven so successful that it has been incorporated into an IBM commercial product.

Information is particularly relevant to management decision making related to competitive activities. Upjohn, for example, recently formed, in response to management's request for assistance in allocating resources relative to competitors' decisions, a Products and Markets Information Research Team that includes representatives from clinical pharmacology and two internal libraries. This team uses extensive commercial databases, together with internal sources, to deliver timely and accurate information in report format to senior management.[19]

An example of an information-focused management process is the monitoring of strategy, or "strategic control."[20] Discussed for at least a decade, the process has received fresh impetus from recent technological changes and business forces. A strategic control system continuously tracks the implementation of strategic initiatives. But few executives who invoke in boardroom discussions the need for faster product development, better customer service and retention, and more and better quality programs also mandate the management processes and enabling systems needed to ensure that these programs are successful. It is these features that differentiate strategic control from traditional management reporting or from an EIS. A complete strategic-control program involves designing performance metrics for tracking strategic progress, building systems for collecting and distributing these metrics, and creating management processes for evaluating reported results. Although few firms today have all the pieces needed to assemble a comprehensive strategic-control system, such systems are likely to be the dominant form of executive information processes in the future.

It is important to recognize that managers not only have input into the design of systems such as these, but are also the chief users. They are involved in every step of the entire process and continuously work with the information the system provides. Such systems are dynamic, with frequently changing metrics, measures, formats, and timing.

At one of Europe's largest firms, Imperial Chemical Industries (ICI), a small number of strategic "milestones," both financial and nonfinancial, are consensually agreed upon. As one company executive noted, by using a small number of success factors ICI can "focus attention on the few really key things underlying the business."[21] Among the nonfinancial measures used by ICI divisions are: product quality levels, process introduction dates, and market share penetration relative to new product launches. The firm's compensation and reward structure, as well as capital allocation and budgeting decisions, are tied to achievements relative to selected milestones.

In summary, for the effective use of information in management processes, there are several important factors to consider. A computerized executive information system will not meet all information needs of executives. Any process should take a wide variety of information sources into account, including those outside the organization. Information must eventually be structured along process lines, rather than functional lines. Finally, it may be helpful to focus on a small set of financial and nonfinancial indicators of business performance.

INFORMATION-ORIENTED OPERATIONAL PROCESSES

Some operational processes are intended primarily to manipulate and generate information. In more and more businesses, for example, the end product is a unit of information—an insurance policy, consulting report, stock transfer, legal brief, advertising campaign, movie, or television program. Even businesses that produce more tangible products have operational processes from which information is the primary output—such support-oriented activities as market and competitor research, customer service, human resource evaluation and compensation, and some financial processes, for example. The management and distribution of employee expertise are other information processes that are beginning to receive attention. But unless these information processes are highly transactional and repetitive, as in banking and insurance, they are unlikely to have been viewed and managed as processes. No one knows where they start or end or how their performance should be measured. Those who perform such activities, usually professionals, are unlikely to view their work in process terms, and often there is no vision of how the process will be performed in the future.

Transaction-oriented information processes (which we treat in later chapters that deal with operational processes), although they create information as an output, often resemble manufacturing processes. The result has been that many transaction-oriented information processes have been measured and improved over many years. Most large banks, for example, can accurately state the unit cost of processing a check; many insurance companies have developed well-oiled "machines" for printing and mailing customized insurance policies. Information processes that involve interorganizational transactions, such as treasury and cash management, money transfer, or large institutional purchases of stock, are also likely to be relatively efficient. Yet in all of these transaction-driven information processes, there remains the potential for radical change that eliminates major steps or functions.

Not all information-oriented operational processes are as highly structured, however. Examples of less-structured information processes can be found in professional services firms and in the management of valuable information such as market research or skills. We discuss both types of examples below.

As they begin to place greater value on the information within their organizations, companies are driven to try to create more formal processes for its management. A few firms have selected such processes for formalization and improvement, if not innovation. IBM's "market information capture" process, one of the company's 18 "enterprise processes," for example, encompasses the management of all information that might lead to the successful sale of products or services to customers. This includes information about products, competitors, customers, market projections and analyses, and even internal skills and capabilities. Customer reactions to products, services, and processes are solicited, captured, analyzed, and distributed to the parts of the organization that can benefit from them by another, related information process termed "customer feedback." Xerox has assigned a somewhat lower level of effort and priority to the process of "intellectual property management." Management of internal knowledge, from skills to patents, although not one of the firm's 14 key processes, is recognized as being of great long-term importance to success.

A number of strategies that address these types of processes are beginning to emerge. We discuss four of them below.

- Because existing information processes are so unstructured, moving to any sort of structured process is itself an innovation. Firms that work on these processes should not worry about order-of-magnitude improvements over the current state, simply because the current state is usually impossible to measure. Their first task, instead, should be to establish an information management process to provide a baseline upon which subsequent efforts can try to improve.

- Information management processes should include the entire information "value chain," that is, the process should start with the definition of information requirements, and move through collection, storage, distribution, receipt, and use of the information (see Figure 4-2). Perhaps the most neglected aspect of the process in most firms is requirements definition, whereby the information users communicate their needs to the provider, whether internal or external. When this exchange is perfunctory, which it often is, the result is usually

Figure 4-2 A Process for Information Management

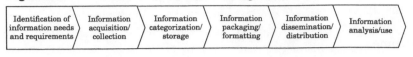

| Identification of information needs and requirements | Information acquisition/ collection | Information categorization/ storage | Information packaging/ formatting | Information dissemination/ distribution | Information analysis/use |

poor-quality information. Also, unless the use of the information is treated as part of the process, the relationship between the information provided and the decisions or actions taken on that information can never be understood or improved. As Peter Drucker has observed, if we are going to create information-based organizations, we must begin to change how people behave with respect to information.[22]

• Because of its power to distribute information throughout a firm, information technology tends to be a key focus of many firms' process efforts. We believe this emphasis is often misplaced. Ironically, because of the unstructured nature of a firm's information needs and flows, IT may be less valuable as an enabler of process change in these types of information processes than in many other types. Navigating the maze of information in a firm almost always requires substantial assistance from human experts, at least with commercially available technologies. Until new technology solves this problem, it is usually better to concentrate on the more human, skill-based aspects of information processes, at least in the early phases of design and improvement.

• Because these processes provide support for some of an organization's core operational processes, their perceived importance ebbs and flows with economic results. When times are hard, the resources needed to support information processes are often cut back, resulting in the loss of skilled people. A process orientation may serve to institutionalize (and measure the economic value of) unstructured information processes, and thereby help to prevent the departure of irreplaceable information skills.

Professional service organizations are coming to realize that information management, when used as a component of a process, can pay substantial dividends. Many consulting firms, for example, are developing systems designed to capture their organizational knowledge and human assets. Such systems enable consultants, attorneys, accountants, or other professionals to better match their firms' institutional and individual skills to a client's problems. One large general consulting firm makes individual professionals' resumes, sorted by skills, project participation, prior work experiences, and educational background, available to all professionals in every office throughout the world. The system is accessed constantly by professionals trying to respond as quickly as possible to client requests for proposals. Operational departments, with full cooperation from the firm's human resource department, work to keep this system current. To ensure compliance, no project can be closed without a full report of who did what and at what level. Other types of systems within the same firm capture copies of all project deliverables, speeches, and internal analyses, and record key content issues in each client engagement.

Many other types of operational processes contain information-oriented subprocesses. In terms of structure and previous process orientation, these tend to lie somewhere between the transaction-based information processes and the unstructured information processes just described. Yet they are critical to process innovation success. For example, a number of firms that have undertaken order management processes have found two types of information subprocesses to be at the core of their efforts. One is the management and distribution of product configuration data, which must be accurately communicated among customers, the sales force, manufacturing, and the logistics functions. Often, no one of these functions actually owns the configuration information subprocess, which means that the potential for error is great. Moreover, to configure, price, and schedule a customer order in real time, frontline personnel must have extremely accurate and accessible information about customers, products, prices, credit policies, shipment schedules, and so forth. Creating such a high-quality information base is beyond the ability of most information systems functions today.

PROCESS INFORMATION MANAGEMENT

Given the importance of information to processes and the importance of information processes to a firm, how should we begin to manage information in a process context? Current methods of information management were inadequate even in the old world of haphazard information requirements and unclear relevance to business goals. Information management practices must be changed radically in this age of information processes and information-based organizations.

A key aspect of information management is the structure of the organizational units that have responsibility for it. For the most part, although information technology has been heavily managed, the management of information, in the sense we have discussed it, has been largely neglected. This is not surprising, given many senior managers' perception that information acquisition, analysis, and distribution is work better managed by subordinates. And although information technology decisions have become sufficiently costly and strategic to force them onto executive agendas, the focus is usually on capital allocation issues and decisions involving internal functional operations. We see little senior management focus on information itself.

The reasons for this situation are complex. Some stem from the perception that information systems and technology are by nature back-office matters best managed by technical people. This leaves information decisions in the hands of either the technical staff or the information-issuing function (e.g., finance). Another reason is that IT has been oversold by vendors, consultants, journalists, and even many firms' own IT professionals as a complete answer to organizational information needs. Shoshana Zuboff put it succinctly: "There is still this childlike belief that technology will make things perfect."[23]

This focus on technology rather than information has led to a backlash against information issues in general. Many executives feel that the payback on the spectacular amount of money spent on IT has been questionable at best. The reaction that has set in against this intricate weave of commerce and enthusiasm has even affected the use of information. Recently, the very notion that information system innovations can confer sustained strategic advantage has been seriously questioned.[24]

It is also clear that the functions responsible for handling unstructured or external information in firms have not succeeded in capitalizing on such information's business potential. These functions, usually known as corporate libraries or information centers, have often been neglected because of their isolation from anything strategic or operationally critical. As evidence of their isolation, a recent survey of executives who had corporate libraries and information centers among their reports found virtually no knowledge of how to value these functions effectively, and little more knowledge of what tasks they actually performed.[25] Because they are rarely perceived as strategic, such functions are often managed as overhead staff functions, and although there may be a strong "feeling" that they are doing something valuable, performance measurement tools to accurately gauge their work do not exist.

There are many examples of successful focused information functions—information management work focused on a specific business need that may cross traditional functional lines. There are also many examples of how these information services, commonly known as "competitive information centers," fill their information niches effectively.[26] Although present and valued in American firms, such functions achieve their apotheosis in Japan. In a number of Japanese companies, the competitive information department collects detailed trip and conference attendance reports from all participants, from middle-management to top-level executives, and refines and distributes these reports to those throughout the firm who the information professionals determine might benefit from them.[27] An attendee at a world industrial ceramics conference, for example, wrote a report that explored how ceramics could be utilized by the firm, which had never before seriously considered using ceramics. The report was sent to the firm's senior management by the company information department, and it was subsequently scrutinized and discussed by the firm's board of directors.

Toshiba's Business Information Center continuously scans all major commercial databases for news about the firm's competitors and business environment. English-language documents identified this way are scanned and translated by computer. This information is further refined and then electronically distributed to more than 600 daily users throughout corporate headquarters, branch offices, and factories.

A recent report on competitive information centers points out that the more such centers get close to key management processes, such as strategy planning, the more likely their chance of organizational success. Centers that remain function-bound and are perceived primarily as passive information services tend to have frustrated staffs and are perceived as doing less effective work.[28]

Although examples of successful information functions do exist, it may be more effective to assign information ownership and stewardship responsibilities to key executives. A few firms have begun to assign information ownership and process ownership to the same executives. At one such firm, a *Fortune* 500 specialty manufacturer, the CEO, concerned about the lack of common information and common uses of information within his organization, created a new role, termed "information czar," for several senior executives. These individuals would set the systems standards concerning what information would be used within specific processes throughout the firm and standardize the processes to be followed in using the information. This initiative proved so successful that executives began volunteering for the role in other processes.

American Express and AT&T are among the firms that have made senior executives responsible for information and given them roles that exist alongside the senior information technology executive role. As most chief information officers are concerned primarily with technology issues, these new information executives must live with such titles as "vice president—information services," which do not adequately describe the nature of the tasks they perform. Examples of these tasks include: establishing firmwide information policies; creating and maintaining information warehouses; coordinating firmwide information acquisitions; initiating efforts to improve the quality of information; building information centers, libraries, or research centers based on user needs; designing firmwide information products and services; and negotiating the sharing of information between divisions.

The skills needed to implement information-focused management processes, as this list suggests, are quite distinct from those employed by most CIOs, suggesting that firms interested in information-focused processes should consider a more "hybrid"-type executive—one with business, technology, and information skills—to lead these initiatives.

Another key aspect of process-oriented information management is its structure or "architecture." Although we are only beginning to research and understand this issue, it is possible that the information architecture to support a process view of business is distinct from nonprocess-oriented architectures. A key focus of the latter is the concept of information engineering—the construction and maintenance of detailed models of data element usage and relationships throughout an entire enterprise. Information engineering is a rigorous approach to separating applications from data, avoiding redundant information storage, and ensuring that data requirements are exhaustively defined. Yet it may not be the correct information paradigm for information-oriented processes.

Although the individual data element is the building block of an information-engineering model, higher-level and more understandable information units may be more appropriate for designing process-oriented information architectures. Because the flow of documents often defines the flow of a business process, we have experimented with the use of the document as the primary unit of information analysis. Rather than laboriously describe the data elements that go into and come out of a process step, one might simply discuss the documents that are required and supplied by the process activity. External market data usually arrive not as a collection of structured data elements, but as a document. Business executives understand the document as a unit of information; they are not usually interested in data elements. Of course, someone in the information technology function must worry about what data elements appear on what documents. But returning to a document-oriented view of information (as was prevalent before the computer era began) means a return to greater simplicity, less detail, and the ability to accommodate less-structured information.

The ultimate goal of information engineering is to create robust, well-structured, nonredundant databases for key "subject areas"—customers, products, and so forth. This is still a worthy, albeit difficult to achieve, goal in a process-oriented environment. But much of the day-to-day work of supplying processes with information is likely to come from other methods and sources. Many firms are finding that so-called composite information systems, which draw from multiple databases to integrate data at the workstation level and can be constructed quickly without rebuilding the entire information infrastructure, are well suited to process information needs.[29] Although some IT function managers are nervous about creating this less-robust information architecture, if processes are to be served with information in the short term, there may be no other choice.

Along with these composite information systems, process information architectures often make use of the information warehouse concept, whereby information useful to processes, be it documents, extracts from subject databases, or small, newly created databases, is stored in easily accessible form for frequent and wide use. Frito-Lay, for example, has used data from its sales transaction systems (which originates at least partially from the sales force's handheld computers) to create marketing information warehouses that enable multiple levels of management to access data for planning and decision-making purposes.[30] Digital Equipment has created a financial data warehouse to support various financial processes. Purists may question the long-term proliferation of information warehouses, but in the short run they are an excellent means of providing processes with information.

One final aspect of information architecture is worthy of mention in a process context. Many experts feel that the successor to information engineering as an information architecture is object-oriented design and analysis. Instead of separating data and applications, object orientation combines data and applications functionality into sets of objects, which communicate with one another by sending messages. The advantages of this approach include greater application simplicity and usability; a "bill the customer" object, for example, might appear in many different systems without having to be rewritten.

Although business applications of the object orientation are just beginning to emerge, some forward-looking firms, believing that objects might be used to combine not only data and applications functions, but also the business rules or subprocesses to be followed in using information and system functions, are already beginning to combine object thinking with process thinking. Thus, if the "bill the customer" object were sent from one division to another, the receiving division would have the same data, same system functionality, and same business process. Opportunities for efficiency and exchange of people and information across processes are obvious.

For cross-functional processes to be supplied with information, managers and professionals must agree on the meaning and usage of information entities, and information must be widely and freely shared across functions and units. In most organizations, this is an unlikely prospect. Moreover, the problem can rarely be solved through technical means. The reason this state can so rarely be achieved is the prevalence of information politics.

Almost every designer of information systems is familiar with information politics. For many reasons, both valid and ignoble, individuals find it in their interest to hoard or attach local interpretations to information. A common reaction to this phenomenon is to decry it and to assume that it cannot be dealt with rationally. A more appropriate reaction would be to identify the political issues, political players, and best strategy for engineering a more appropriate political response. In a recently completed research project on information politics, we discovered that some models of information politics, including feudalism, anarchy, and technological utopia, were much less effective at sharing information across processes than information monarchies and federalist states.[31]

SUMMARY

This chapter has dealt in an exploratory fashion with the relationship between processes and information. We have considered the role that information plays in improving and innovating processes, including performance monitoring, integration across and within processes, and process output customization. We also discussed various types of information-oriented processes, both management and operational. Finally, several aspects of appropriate information management in a process context were presented. Issues of organizational structures for information management, information architectures, and information politics were considered.

Although information and information technology together constitute a formidable enabler of process change, they are not the only factors that can lead to radical process change. People—both as individuals and in organizations—are also a major enabler of change in how work is done. In the next chapter, we discuss the role of human and organizational enablers of process innovation.

Notes

[1]Blaise Cronin, ed., *Information Management: From Strategies to Action* (London: Aslib, 1985).

[2]For a sampler from these fields, see Fritz Machlup, *The Study of Information: Interdisciplinary Messages* (New York: John Wiley, 1984).

[3]Joseph M. Juran, *Juran on Planning for Quality* (New York: Free Press, 1988): 97.

[4]David Garvin, *Managing Quality* (New York: Free Press, 1988): 169.

[5]Ibid., 204–207.

[6]Shoshana Zuboff, *In the Age of the Smart Machine: The Future of Work and Power* (New York: Basic Books, 1988): 9.

[7]James P. Womack, Daniel T. Jones, and Daniel Roos, *The Machine That Changed the World* (New York: Rawson Associates, 1990).

[8]Steven Dichter, "The Organization of the 90's," *McKinsey Quarterly* 1 (1991): 145–155.

[9]Sharon M. McKinnon and William J. Bruns, Jr., *The Information Mosaic* (Boston: Harvard Business School Press, 1992).

[10]See, for example, John F. Rockart and James E. Short, "Information Technology and the New Organization: Towards More Effective Management of Interdependence," working paper 90s:88-058, MIT Sloan School of Management, Management in the 1990s, September 1988.

[11]"Who's Being Helped by Help Lines," *New York Times* (May 14, 1991): D4.

[12]For a broader discussion of market information management, see Vincent P. Barabba and Gerald Zaltman, *Hearing the Voice of the Market* (Boston: Harvard Business School Press, 1991).

[13]For a detailed analysis of the mass customization concept, see B. Joseph Pine II, *Mass Customization* (Boston: Harvard Business School Press, 1992).

[14]See, for example, H. Thomas Johnson and Robert S. Kaplan, *Relevance Lost: The Rise and Fall of Management Accounting* (Boston: Harvard Business School Press, 1987).

[15]*The Management Challenge of Information Technology* (London: Economist Intelligence Unit, 1991): 86.

[16]McKinnon and Bruns, *The Information Mosaic,* 105.

[17]Frank Aguillar, *Scanning the Business Environment* (New York: Macmillan, 1967): vii.

[18]Personal interview with IBM Latin America executives.

[19]Jan Dommer, "Maintaining the Competitive Edge at the Upjohn Company," *Inside Business* (Spring/Summer 1991): 13.

[20]See, for example, Michael Goold and John J. Quinn, *Strategic Control: Milestones for Long-Term Performance* (London: Economist Books and Hutchinson, 1990).

[21]Ibid., 41.

[22]Peter F. Drucker, "Viewpoint: What Executives Need to Learn," *Prism* (Cambridge, Mass.: Arthur D. Little, 4th Quarter 1990): 73–84.

[23]Ronald Henkoff, "Make Your Office More Productive," *Fortune* (February 25, 1991): 73.

[24]Max Hopper, "Rattling SABRE—New Ways to Compete on Information," *Harvard Business Review* (May–June 1990): 118–125.

[25]Laurence Prusak and James Matarazzo, *Valuing Corporate Libraries* (Washington, D. C.: Special Libraries Association, 1989).

[26]For a good example of a well-functioning business intelligence group, see Jane Linder and Margaret King, "Digital Equipment Corporation: Leadership in Corporate Intelligence," N9-192-002. Boston: Harvard Business School, 1992.

[27]James Matarazzo and Laurence Prusak, *Japanese Success and Information Management* (Washington, D.C.: Special Libraries Association, 1991).

[28]Diane Weston, "Best Practices in Competitive Analysis," *SRI International Report* 8D1 (Spring 1991).

[29]Stuart Madnick and R. Wang, "Evolution Towards Strategic Applications of Data

Bases Through Composite Information Systems," *Journal of MIS* 5:2 (Fall 1988): 5–22.

[30]See Barnaby J. Feder, "Frito-Lay's Speedy Data Network," *New York Times* (November 8, 1990): D1; also Nicole Wishart and Lynda Applegate, "Frito-Lay, Inc.: HHC Project Follow-Up," 9-190-191. Boston: Harvard Business School, 1990.

[31]For more information on this topic, see Thomas H. Davenport, Robert G. Eccles, and Lawrence Prusak, "Information Politics," *Sloan Management Review* (Fall 1992).

Chapter 5

Organizational and Human Resource Enablers of Process Change

To focus only on information and associated technologies as vehicles for process change is to overlook other factors that are at least as powerful, namely, organizational structure and human resource policy. In fact, information and IT are rarely sufficient to bring about process change; most process innovations are enabled by a combination of IT, information, and organizational/human resource changes.

Attention to both social and technical factors as agents of change arises from a long tradition, having been a focus of the "sociotechnical systems" approach to understanding and managing change developed at the Tavistock Institute of Human Relations in London in the 1950s.[1] Although information technology was not part of the initial focus of the research, Enid Mumford and others demonstrated that sociotechnical systems principles apply to information as readily as to other types of technology.[2]

The sociotechnical theorists did not have the advantage of a process orientation, with its customer and measurement focus, and they typically made no distinction between incremental and radical levels of change. Sociotechnical plans also were not firmly linked to strategy and operational vision. Information technology did not even exist as a useful organizational tool for much of the period of sociotechnical research. Still, the sociotechnical approach has taught us many useful lessons, which we call attention to at relevant points throughout this chapter.

Due in part to the pioneering efforts of the sociotechnical school, the changes in organizational structure, behavior, and policy that enable process innovation may not be as innovative in an absolute sense as those derived from IT. But organizational structures and human resource policies need not be entirely new

95

to be effective as change enablers, only untried within a specific organization or process.

Because they have been a part of the enterprise for a much longer period of time, organizational structure and human resource policy are more familiar to managers as change tools (although managers have not universally mastered their use). The great irony is that familiarity seems to have bred neglect, in part because the evangelists of process innovation are much more likely to lead the information services function than the human resource function. They undertake carefully managed projects, employing tested methodologies and strict timetables, to build new systems enabling processes that, because the human aspects of change are managed as afterthoughts, lead to significant human resource problems.[3] Too many systems fail to yield any real business benefit because of human problems in implementation.[4]

If process innovation is to succeed, the human side of change cannot be left to manage itself. Organizational and human resource issues are more central than technology issues to the behavioral changes that must occur within a process.

If sociotechnical research has taught us anything, it is that, for process innovation to succeed, all enablers must be aligned and in balance with other key aspects of the organization. If, for example, the technological innovations in a process enable greater worker empowerment and autonomy, the organizational culture must be adjusted to support those directions. Conversely, if an organization's culture supports control and maximum efficiency, systems to enable process innovation must be consistent with these objectives to succeed. We are not enthusiastic about control-oriented cultures (we wouldn't want to work in such environments), but they are possible and sometimes necessary. In this chapter we discuss organizational and human resource changes that support both empowerment and control, and, as we did for technological enablers, explore both the opportunities presented and constraints imposed by these enablers.

ORGANIZATIONAL ENABLERS: STRUCTURAL AND CULTURAL

Organizational enablers of process innovation fall into two categories: structure and culture.

Structural Enablers: Team Approaches

Of the many kinds of structural changes that can facilitate new, process-oriented behaviors, one of the most powerful involves structuring process performance by teams.

Since work design was first advanced during the Taylor era, the primary unit of work performance has been the individual. Taylor and others believed that the more isolated the worker, the more efficient the performance of the task. Yet most processes or subprocesses can be performed by teams (or collections of teams). Work teams were studied in terms of the productivity, motivation, and satisfaction of workers by primarily Harvard researchers at the Hawthorne plant of Western Electric as early as the 1930s,[5] and analysis of the role of work teams in relation to new technologies and work processes was pioneered by the Tavistock School in the late 1940s.[6] The technologies involved in the latter research were not computers and communications, but new machines for coal mining. Tavistock researchers found that individual-based work designs combined with new technologies were less productive than work teams with no technology. The rationale for this finding was, to a degree, process-oriented; the teams worked better because they combined multiple functions into one unit, allowing adaptation to changing conditions inside the mine. Subsequent studies of work teams in a variety of settings—white and blue collar, on assembly lines and in less-structured contexts—have found significant benefits to be associated with a shift to teams.[7]

Benefits of teams.   Notwithstanding this history of team experiments and analysis, American firms have, until recently, been slow to adopt team approaches on a large scale. Firms that have begun to explore the use of autonomous teams as the primary unit of work organization—General Electric, Xerox, Martin Marietta, Aetna Life Insurance, and others—are seeking specific benefits from doing so.[8]

First, they are looking for cross-functional skills in single work units. Cross-functional skills facilitate functional interfaces and parallel design activities. Furthermore, a broad set of skills and perspectives increases the likelihood that output will meet multifunctional requirements. New product development teams, for example, increasingly include representatives from all the functions involved in the product development process. Even teams composed of a number of employees from the same function are likely to have a broader range of skills than any individual. Nonaka has cited the wealth of information created through overlapping roles and memberships on teams as a key source of innovation in Japanese product development.[9]

A second benefit sought by companies that employ team structures is improved quality of work life. Most human beings seem to prefer jobs that include social interaction, and work teams provide opportunities for small talk, development of friendships, and empathic reactions from other employees. Such socialization was anathema to Frederick Taylor, but although there may be a trade-off between productivity and socialization on the job, it is also true that alienated, unhappy individual workers are no more productive than overly socialized team workers. Our experience with companies that have adopted team structures has been that productivity has either remained stable or increased. It is well known that team structures are common in Japan, whose companies are noted for their productivity levels; process improvement and quality control in Japan, according to Ishikawa, "is a group activity and cannot be done by individuals. It calls for teamwork."[10]

The socialization benefit is particularly important when the primary content of the work is informational. The most efficient process designs for such work can chain workers to terminals. With no team structure to foster social interaction, the workers performing these job tasks can become alienated. Zuboff has observed this tendency in a number of case studies, particularly in financial services industries.[11] The Internal Revenue Service, for example, introduced a new system for managing the work of collections agents that greatly simplified the process, and considerably increased collections, with half the headcount. But workers considered their jobs to have worsened markedly. Instead of socializing with co-workers as they discussed or sought taxpayer case files, they worked "heads-down" at terminals with telephone headsets, stopping only for breaks. The system was an economic success but a work life failure; collections agents resigned in large numbers. When a redesigned process gave cases to self-managing teams rather than individual workers, the level of job satisfaction increased greatly with no loss in economic benefits.[12]

It should also be noted that the social interactions of team members are not always positive. Particularly when teams are cross-functional, members may lack a shared culture, leading to conflict and misunderstanding. This conflict has been noted, for example, in cross-functional teams for new product development,[13] and in teams for creating integration across the design and manufacturing functions.[14] In some cases, these conflicts have significantly decreased the innovativeness and performance of the team. Therefore, careful attention must be paid to cultural compatibility issues in selection of team members. Ancona and Caldwell suggest, for example, that matching team members in terms of their tenure in the organization can help to counter the negative effects of functional diversity.

Types of teams.   Several different types of teams are relevant to process innovation. Cross-functional teams employed to design processes, that is, in the "process design process," are discussed in a later chapter. Our focus here is on teams organized to execute work, which is typically longer in duration than the process design activity, even indefinite. This time frame, as discovered in an analysis of 27 diverse teams, gives rise to a set of dynamics that is not present in short-term teams or task forces.[15]

The most difficult issues that face teams organized to execute processes or subprocesses over the long term revolve around the relationship between team members and the functional structure of the organization. Assuming the functional structure continues to exist, as it has in every organization we have studied, one of these issues is the question of how and by whom the team members are to be evaluated. Federal Mogul, an auto parts company, created business unit teams to speed the process of product prototype development. The firm established teams whose members had both process team and functional responsibilities, but who continued to be evaluated by their functional superiors. This design introduced potential sources of conflict with regard to process versus functional emphasis. A Canadian high-technology manufacturer heavily involved in process design efforts experienced conflict between process and functional emphases even in its design teams. "As long as the members of these design teams are evaluated by their functional managers," observed one manager, "their willingness to suggest process designs that weaken or reduce headcount in functions will be compromised."

One obvious solution to the problem of emphasis is to create a process-based organization, whether stand-alone or one that works in conjunction with the functional organization, and give process representatives a role equal to that of functional managers in evaluating and compensating team members. This step should, however, be taken well after process innovation initiatives are under way.

Issues of team formation and function involve management as well as operational employees. In fact, a number of the CEOs we interviewed remarked that one of their most important and difficult tasks was to create a cross-functional team to direct and oversee process innovation efforts. "The hardest task in redesigning our processes," explained the managing director of a European office products company, "was to get our senior managers to work as a team. They had to take off their functional hats and put on their company hats in order to think cross-functionally. A couple of the managers never made the transition, and they had to leave."

To function effectively as a cross-functional team, a senior management group must be willing and able to look beyond functional allegiances, and even beyond what may benefit their careers. That many today arrive at management positions by way of ambition and political skill does not bode well for team formation or success. But this is not to suggest that what we need is egoless managers; the Western version of management teams must combine a group orientation with the freedom to think independently and creatively.

Finally, although we have distinguished between work execution and process design teams, it is important to recognize that the former can usefully be assigned responsibility identifying opportunities for process improvement. In fact, if improvement is to become part of the fabric of work in a team-oriented work environment, team structures are the right unit for streamlining or quality efforts.

Criteria for team success.   Whatever the type of team, one can find research and experience that indicate how it can be successful, both in general and relative to process innovation in particular. Composition, for example, has been shown to be key to team success. Although personality issues may render some productive workers unsuited for teamwork, in general, employees who perform well as individuals tend to perform well as team members.

Finding good performers will contribute to productivity, but unless, collectively, they possess the functional background, skills, and experience requisite to the process being innovated, the team is not likely to succeed.

A clear relationship to functional structure—in terms not only of reporting relationships and performance and reward evaluation, but the relative emphases on process versus functional activities—is also key to team success, particularly if a team is temporary. We have been told by process design team members that it is difficult to allocate attention to teamwork unless it comprises at least 50% of their jobs. In other words, a team that meets only one day per week is not likely to succeed.

Logistical issues, such as the location of team members, can also affect team success. In one computer company that was redesigning several key processes, the owners of each process themselves formed a team. A division in which process owners were colocated, enabling them to meet biweekly, progressed much faster in designing new processes than other divisions in which geographical separation precluded such frequent meetings.

It is popular today to establish "self-managing" teams, teams that direct their own work and have no formal leader.[16] Although desirable because of its positive effect on motivation,[17] self-management can create ambiguity about who really manages a set of activities. Davis-Sacks, for example, describes a team of highly competent professionals in the U.S. federal government; notwithstanding its creator's intention that it be self-managing, the team's leader believed herself to be in charge and insisted that all communications go through her.[18] The team's failure to complete its assigned task on time and, ultimately, to have any effect on the process it was created to facilitate was shown to be a result of this lack of clarity in team management.

One aspect of team success that varies with process type and industry is boundary management. The need goes both ways; a focus on intragroup effectiveness and dynamics must be combined with attention to a team's relationship with the larger organization. At the same time, the larger organization must support the team with adequate resources, including management attention when needed. Process teams that interface directly with customers—typically a stressful relationship—must maintain two boundaries: the team may satisfy the customer while antagonizing the rest of the firm, or vice versa.

In the many examples of case management we have observed, in which the whole of order management or some other customer-facing process is handled by a team (or individual), the team (or individual) is usually empowered, at least in principle, to satisfy customers and resolve their problems. Given the company's need to profit from sales or service transactions, case managers need to have a clear understanding of how far they can go to satisfy the customer.

Clarity—in mission, process boundaries, decision-making authority, and internal and external roles—emerges as the single most important success factor for process teams. Clarity is facilitated when teamwork is viewed in process terms; the process orientation supplies a clear purpose (producing the process output), and the process performance objectives become the performance objectives of the team.

Finally, it must be acknowledged that teams are not the solution to all process design problems. The work of individuals can also be effective, and Taylor's concern about the efficiency of teams is not unfounded. Hackman noted:

> Teams and work groups have a shady side, at least as they typically are designed and managed in contemporary organizations. They can, for example, waste the time and energy of members, rather than use them well. They can enforce norms of low rather than high productivity. They sometimes make notoriously bad decisions. Patterns of destructive conflict can arise, both within and between groups. And groups can exploit, stress, and frustrate their members—sometimes all at the same time.[19]

As a result, teams must be carefully designed, and alternatives to teams considered. The design of work teams is an extremely complex subject, about which there is a vast literature. Creating a team-oriented approach to process execution should not be undertaken without assistance from experts. Furthermore, organizations should not adopt team structures for every process environment. Teams are a means of building social interaction and a cross-functional perspective into a work process. If these objectives can be accomplished through some other means, or are for some reason unnecessary, team structures are not required.

Team-oriented information technologies.   Emphasizing the synergistic relationship between technological and human/organizational enablers of processes, there are emerging technologies to facilitate work in teams. The technologies that support team structures have spawned wholly new forms of software. One of these, termed "groupware," supports a variety of activities intended to facilitate group work, including:

- group brainstorming, decision making, and structured discussion;[20]
- group communication via teleconferencing, electronic mail, and electronic bulletin boards;
- group preparation of documents;
- group scheduling of meetings and facilities;
- group access to database records (e.g., customer files); and
- analysis of group processes (e.g., frequency of participation).

According to one source, these types of groupware functions have many potential benefits, including reducing meeting time and face-to-face meeting frequency, curtailing missed communications, reducing project delays, and enabling increased flow of information and faster decision making. In Balcor, a real estate investment management firm, for example, these benefits added up to a 30% reduction in product development time from the use of groupware.[21] However, neither these researchers nor Johansen and Bullen,[22] nor various researchers in Greif,[23] treat the use of groupware specifically within a process context. Though there is no research to support conjectures on groupware use in team-based processes, we would suspect that the more structured the process, the more likely groupware use will lead to measurable time reductions or other benefits.

Other technologies that do not traditionally fall under the groupware rubric can also facilitate the work of process teams. Team-based processes that involve telephone work often take advantage of automatic call distributors (ACDs) that route incoming calls to individuals who are free or whose expertise matches customer keypad responses. At the Internal Revenue Service, however, when the collections function attempted to move from individual work to self-managing teams, one of the most difficult aspects of the change was to enable teams to override the movement of calls and cases as directed by the ACD and a computer system.

So-called work flow software, which can route documents much as ACDs route telephone calls, is helpful in environments where imaging has replaced paper as the primary means of transferring work among team members.[24] Routing can be based on employee availability or case difficulty, and can be changed to accommodate employment shifts or vacations. But although imaging vendors promote the ability of work flow software to enable radical shifts in the flow of work within teams, few companies have deviated substantially from the path blazed by paper.

Computer-supported process teams remain relatively rare, but it is clear that computers can help teams structure work and track customer or case progress through a process, and facilitate intra- and interteam communications. As these support technologies mature, discretion in the use of their monitoring and tracking capabilities will become increasingly important; excessive use of computer controls could easily neutralize the motivational advantages of self-managing teams.

CULTURAL ENABLERS

Most recent shifts in organizational culture have been in the direction of greater empowerment and participation in decision making and more open, less hierarchical communications. The resulting participative cultures, which have a structural side in flatter organizational hierarchies or broader spans of control, have been widely documented to lead to both higher productivity and greater employee satisfaction.[25]

In a process innovation context, these cultural changes are intended to empower process participants to make decisions about process operations. Participative cultures may even lead to self-design of smaller, restricted processes by employee teams.

Customer-facing processes such as order management and customer service are, as we shall see later, well suited to empowering frontline employees to satisfy customer demands. The three enablers of teams, empowered team members, and more information on customers can yield radical improvements in customer-oriented processes. These improvements are the heart of the case management movement discussed later.

Worker empowerment is also related to the general innovativeness of the corporate culture; 3M Corporation, for example, has long been renowned for its climate of innovation and its success in developing new products and processes. "As our business grows," explained 3M president William McKnight in 1944,

> it becomes increasingly necessary to delegate responsibility and to encourage men and women to exercise their initiative. This requires considerable tolerance. Those men and women to whom we delegate authority and responsibility, if they are good people, are going to want to do their jobs in their own way. Mistakes will be made. But if a person is essentially right, the mistakes he or she makes are not as serious in the long run as the mistakes management will make if it undertakes to tell those in authority exactly how they must do their jobs. Management that is destructively critical when mistakes are made kills initiative.[26]

Although process innovation is not normally a bottom-up activity, a culture that is receptive to innovation at all levels is likely to both identify and implement process innovation at relatively high frequencies. Furthermore, even after broad process designs have been implemented, an innovative culture can inspire minor improvements that benefit day-to-day process performance.

By no means are all organizations moving toward greater degrees of empowerment—nor is it necessarily appropriate that they do so. Processes that involve largely menial work performed by low-skill, high-turnover employees not expected to be committed to their jobs may be more appropriately executed in a control-oriented culture. Control in such environments ensures the quality and efficiency of the work and guarantees that the knowledge does not reside solely with employees.

Information technology, as we pointed out earlier, can support either culture—control or empowerment. It can supply employees with information that enables them to make their own process decisions or with instructions that dictate precisely how to perform each process step. This dual nature of IT, which enables individuals or teams to manage themselves with information, or information to be used to closely monitor individual or team performance, has been noted by Zuboff[27] (whose terms for it are "informate" and "automate") and Walton[28] (who uses the terms "compliance effects" and "commitment effects").

Control-oriented process cultures are most frequently seen in the service industries; fast-food and lodging are good examples. Low levels of employee commitment, slim profit margins, and the need for consistency and quality make a controlling culture likely if not inevitable. Firms in these industries constantly seek IT innovations to afford greater control and enable the capture and display of process knowledge. One hotel chain, for example, is planning to use in-room television to display step-by-step instructions for cleaning and preparing rooms. McDonald's' in-store computers plan, monitor, and control many aspects of store operations, for both employees and managers,[29] and Mrs. Fields Cookies relies extensively on information technology to control key processes (although the culture includes elements of empowerment as well).[30]

ORGANIZATIONAL CONSTRAINTS TO PROCESS INNOVATION

Organizational enablers, like IT enablers, have a flip side; they can constrain, as well as enable, process innovation. Most organizational constraints are simply the reverse of the enablers we discussed above. They involve an organizational design element that does not fit the process design, for example, a process design that assumes a great degree of frontline empowerment in a control-oriented organizational structure and culture. But some organizational constraints are more general, for example, organizational structures and cultures that support management along individual functions rather than cross-functional processes. The senior executives who observed that their greatest challenge in moving to process management was getting the senior management group to act as a team were speaking of a constraint to multiple types of change, not just to process innovation.

Many of the firms beginning to investigate process innovation are constrained by some aspect of their structures or cultures, which makes it difficult for them to pursue their interest to the point of concerted action. Structural and cultural constraints to process innovation include strict hierarchical structures, cultures unreceptive to innovation, and general organizational rigidity or inability to accommodate change. These constraints are not all hallmarks of failure; in fact, they may be generated by success. Management and removal of these types of constraints, which must occur before or concurrently with process innovation initiatives if the initiatives themselves are to succeed, are discussed later in this book.

HUMAN RESOURCE ENABLERS OF PROCESS INNOVATION

Though organizational and human resource enablers of process innovation are often inextricably linked, here we focus on enablers that are closely tied to the way individual workers are trained, motivated, compensated, evaluated, and so forth, rather than how their work is structured in terms of organizations and groups. In particular, we focus on skills, job motivation, and human resource policies.

New processes invariably involve new skills. Because process innovation often involves both greater worker empowerment and a broader set of work tasks, the requisite new skills may involve both greater depth of job knowledge and greater breadth of task expertise. A worker expected to be a generalist and participate on an autonomous team, for example, must learn about the jobs of the other team members (cross-training) and, if new technologies are to be employed in the process, must acquire skills in applying and using those technologies.

A variety of training programs must be undertaken if the requisite skills are to be available when they are needed. These include specific process training, anticipatory training, and on-the-job training.

The most common type of training in a process innovation context is specific process training. When a new process is designed, the process skill requirements must be assessed and workers who will execute the process must be trained in those skills. Though it may seem straightforward, there are several problems with this type of training. One, because skill acquisition often takes longer than process design, the need for workers with new skills may be urgent before their training can be completed. Two, there are likely to be few employees with sufficient experience and knowledge to train others. If the process is truly innovative, no one will be qualified to train anyone else. Three, unless workers are overqualified for their jobs, it may be difficult to find employees with enough raw intelligence and generic job skills to execute the new process.

Process innovation in Mutual Benefit Life's underwriting processes, for example, created new case management jobs that placed heavy demands on worker skills. Many of those employed in the old underwriting process were not qualified to become case managers, and probably would never be qualified even with extensive training. Because the firm had a long tradition of hiring and training less-educated workers from the local community, the displaced workers felt betrayed. Although the process was eventually made to work effectively and a sufficient number of case managers were identified, these innovations were achieved at considerable cost to the company's paternalistic culture and employee morale.[31]

The problems inherent in specific process training have led a number of companies to undertake anticipatory training aimed at better preparing managers and workers for process innovation or the IT and other enablers associated with it. A number of manufacturing firms, for example, have established education programs that deal with the ideas behind concurrent engineering, design for manufacturing, integrated logistics, and so forth. Although these programs are frequently offered in association with quality management, their message is often radical rather than incremental change.

Other large companies such as Du Pont and Aetna have undertaken broad programs to familiarize senior and middle managers with the capabilities of information technology. Although not created in an explicit process innovation context, these programs have been cited by those who sponsored and delivered them as having stimulated IT enablement of process innovation. Of course, process innovation is not a simple by-product of this education; many other factors must be present. In Japan, broad worker training in various aspects of manufacturing processes has been discussed as an enabler of so-called flexible manufacturing systems, which are certainly a process innovation.[32] An overall organizational culture that emphasizes skill enhancement and job rotation has also been cited as a key factor in the flexibility and innovativeness inherent in Japanese manufacturing processes.[33]

On-the-job training, long known and appreciated, has sometimes been used as an excuse for neglecting other types of training. In the context of process innovation, on-the-job training takes two forms. One of these relates to the notion of process prototyping, discussed later as a means of refining process designs. Process prototyping also has application to the acquisition of process skills. If there is no expert to train workers in the execution of a process, the obvious alternative is to teach the process and its associated skills to a number of flexible, adaptive employees and have them impart their new knowledge and skills to others.

Another approach to on-the-job process skills training, because it makes use of workstations, is particularly relevant to IT-enabled process innovation.[34] The value of this approach is that it enables complex educational materials to be delivered when they are required by the process worker. IBM's internal information-systems function, for example, is attempting to speed processes for developing new computer systems by embedding interactive training materials in the systems developers' workstations. Programmers who encounter aspects of a problem or tool that they do not understand can call up a quick training session that employs videodisk images.

Federal Express has institutionalized the latter method of on-the-job training; it has installed more than 1,000 interactive video workstations for training of frontline couriers and customer service personnel. The workstations enable rapid learning of new products and services as they are rolled out. When, for example, Federal initiates international package delivery service to a new country, it uses the training systems to acquaint employees with customs requirements for the country. The company, viewing the systems as a strategic advantage, has integrated them into many of its human resource policies. Under its policy of learning-based compensation, for example, when a Federal Express employee completes a training module, the workstation automatically increases the employee's compensation level in the payroll database.[35]

In extreme cases, process innovation may demand not just new skills, but new employees. Many process designs lower employment requirements, enabling firms to eliminate all or most workers in the process area and then hire fewer, new employees with different backgrounds and different skills. Although the hiring of new employees as an enabler of process innovation is a viable option, it is not a particularly desirable option for a company that cares about employee loyalty and morale. A better alternative is to hire capable, flexible employees from the beginning and invest to keep their skills current and adaptable to the needs of new processes.

Motivation levels of employees are a key determinant of process performance. Employee motivation results from a combination of factors, some determined by individuals' personalities. Companies can set as an objective the hiring of workers with high motivation levels, but they can also design motivation into their processes. The consensus model in studies of work organization suggests that work motivation derives from five key aspects of the job, or process:[36]

- skill variety (the variety of skills necessary to complete the job),
- task identity (the degree to which a job involves completion of an entire activity),
- task significance (the perceived importance and impact of the job),
- autonomy (the freedom and discretion with which the job is performed), and
- feedback (the extent to which information about the performance of the job is provided to the worker).

Although a process innovation orientation does not guarantee job motivation, a number of aspects of process thinking make some of these motivating characteristics more likely. The cross-functional nature of processes, for example, implies a greater likelihood of skill variety in a particular job. Similarly, the strong output orientation of process-based work organization increases the probability of task identity. And finally, the measurement focus on processes provides a potential source of job feedback. In short, well-designed process jobs are likely to be highly motivating, and high motivation is a key aspect of process performance.

A number of other human resource policies can be viewed as process innovation enablers when combined with technological and other organizational changes. These policies, and their implications for process innovation, are discussed briefly below.

Compensation. Since the 1930s, when Joseph Scanlon began to advocate the sharing of productivity gains in the steel mill where he was a labor leader, companies have experimented with approaches to compensating workers on the basis of operational improvements.[37] This approach, now usually called "gainsharing," is widely referenced in the quality literature. In general, researchers have discovered that compensating workers on the basis of performance leads to productivity increases, although the issues involved are complex.[38] Obviously, given the strong measurement orientation of a process approach, it would be relatively straightforward to compensate process workers on the basis of process performance. Just as obviously, doing so would probably be an effective motivational technique for these employees. At the management or process owner level, compensation-oriented enablers might include giving managers a financial stake in the performance of the process, perhaps even ownership in the literal sense. We are acquainted with several firms, among them Royal Trust (a Canadian bank), AT&T, and Xerox, that are beginning to explore this type of approach in the context of a process innovation initiative.

Career paths.   A process view of the organization usually implies career paths different from those found in the typical functional, hierarchical organization. Career movement is likely to be more lateral than upward; titles may no longer reflect the importance of the role. We are aware of several firms in which new processes already seem to involve fewer options for upward advancement. Companies that can figure out how to motivate employees under these new career conditions will have a long-term advantage in process innovation over those that cannot.

Work role rotation.   Because processes are typically collections of functions, a process worker should know as much as possible about other functions and activities in order to be able to effectively integrate across them. One way to ensure broad process knowledge is to rotate workers through the various jobs in the process or in related processes. This, like process-oriented career paths, is a long-term enabler of process innovation that should be established broadly throughout a company rather than for a specific process.

Lifetime employment.   Although seemingly impossible for most American firms to achieve, a lifetime employment policy greatly facilitates process innovation. Employees who feel that they have a job for life are much less worried about designing or performing their jobs out of existence. IBM is one of the few large companies headquartered in the United States that still maintains a lifetime employment policy (and its policy is weakening rapidly under economic pressures), but many firms around the world do so, particularly in Japan. Large Japanese firms' employment of this policy is often credited for the process flexibility and grass-roots innovativeness of their employees.[39]

Because of their broad and long-term nature, most of the policies described above should be viewed as contextual factors for enablement as much as enablers. They make process innovation through other enablers much more likely to happen. Only the most dramatic change in human resource policies could itself be viewed as a lever for a new process design.

The absence of these human resource enablers could be considered constraints to process innovation. In fact, most change management methodologies of which we are aware focus on identifying and removing organizational and human resource constraints.[40] The positive tone of enablement may spur more enthusiasm for innovation.

MANAGING HUMAN RESOURCES AND ORGANIZATIONAL CHANGE

The primary lesson to be learned from extensive sociotechnical research is that social and technical change must be managed jointly. Any approach to embedding the enablers described in this and the previous chapter in new processes should be a joint method, with concurrent consideration of all the enablers necessary for a specific process design.

Richard Walton has proposed that all major systems changes be accompanied by an "organizational impact statement" that lays out the simultaneous development of the information system and the organization.[41] Although he offers other options—"anticipatory development of the organization" (which may be appropriate for some of the broader and longer-term human enablers described above) and "reactive development of the organization"—he argues most strongly for simultaneous development.

What form simultaneous development should take depends on how an organization plans to implement technical and organizational change. For example, a firm using a systems-development lifecycle methodology to structure the development of new information systems that support a process should manage human change along with the systems change. The company should plan the organizational and human resource changes as it plans the system, design the human change in detail as it designs the system, and construct the new organizational structures, cultures, and human resource procedures as it constructs the system.[42]

A prototyping approach is generally more appropriate than a lifecycle model for implementing process innovations. At the same time that information systems are prototyped using quickly built systems on personal computers, organizational enablers, including skills, motivational approaches, human resource policies, and even, to the degree possible, organizational structure and culture, should be prototyped with a small group of process workers and process inputs and outputs.

The structure of the prototype should reflect two potentially conflicting objectives: to prove that the process design and the enablers work, and to learn from mistakes and missteps made during the "experiment." The conflict between these objectives is readily apparent when the issue of personnel selection is raised. To prove the concept, one would want to choose the best people available. To learn from the prototype, the appropriate people would be of average capability. Companies must strike a balance between these two objectives in pursuing the prototype-based approach to implementation.

SUMMARY

This chapter has considered the role of organizational and human resource factors as enablers of process innovation. Acting alone and in concert with information and information technology, these approaches can lead to radical change in work structure, motivation, and process performance. Like other enablers, human and organizational aspects of the organization can either constrain or provide opportunities for innovation.

The three key enablers of process change—information technology, information, and organizational/human issues—have now been described. As noted in our overview of the approach to process innovation, these enablers should be considered early in the life of a process innovation initiative. After some enablers have been identified as relevant and explored in a preliminary fashion, the organization can begin to construct a vision for the new process, as described in the following chapter. The selected enablers become components of the overall vision of how work is to be done.

Notes

[1]See, for example, A. K. Rice, *Productivity and Social Organization: The Ahmedabad Experiment* (London: Tavistock Publications, 1958), and F. E. Emery, "Characteristics of Socio-Technical Systems," document T176 (London: Tavistock Institute of Human Relations, 1959).

[2]See, for example, Enid Mumford, "Designing Human Systems," (Manchester, U. K.: Manchester Business School, 1983); also by Enid Mumford and D. Hensall, "A Participative Approach to Computer Systems Design," (London: Associated Business Press, 1979).

[3]Cyrus F. Gibson and Thomas H. Davenport, "Systems Change: Managing Organizational and Behavioral Impact," *Information Strategy: The Executive's Journal* 2:1 (Fall 1985): 23–27.

[4]Cyrus F. Gibson et. al.,"Strategies for Making an Information System Fit Your Organization," *Management Review* (January 1984): 8–14.

[5]Fritz J. Roethlisberger and William J. Dickson, *Management and the Worker* (Cambridge, Mass.: Harvard University Press, 1939).

[6]See Eric Trist and K. Bamforth, "Some Social and Psychological Consequences of the Longwall Method of Goal Getting," *Human Relations* 4 (February 1951): 3–38; also see Eric Trist et al., *Organizational Choice* (London: Tavistock Publications, 1963).

[7]Thomas G. Cummings and Edmond S. Molloy, *Improving Productivity and the Quality of Work Life* (New York: Praeger, 1977): 38–49.

[8]Dennis Kinlaw, *Developing Superior Work Teams* (Lexington, Mass.: Lexington Books, 1991).

[9]Ikujiro Nonaka, "Redundant, Overlapping Organization: A Japanese Approach to Innovation," *California Management Review* 32:3 (Spring 1990): 27–38.

[10]Kaoru Ishikawa, *What Is Total Quality Control? The Japanese Way* (Englewood Cliffs, N.J.: Prentice-Hall, 1985): 89.

[11]See Shoshana Zuboff, *In the Age of the Smart Machine: The Future of Work and Power* (New York: Basic Books, 1988): 437, ff. 55.

[12]Nitin Nohria and J. Chalykoff, "Internal Revenue Service: Automated Collection System," 9-490-042. Boston: Harvard Business School, 1990.

[13]Deborah Gladstein Ancona and David E. Caldwell, "Cross-Functional Teams: Blessing or Curse for New Product Development?" in Thomas A. Kochan and Michael Useem, eds., *Transforming Organizations* (New York: Oxford University Press, 1992): 154–166.

[14]Donna B. Stoddard, *Information Technology and Design/Manufacturing Integration*, Ph.D. diss., Harvard Business School, 1991.

[15]J. Richard Hackman, ed., *Groups That Work (and Those That Don't)* (San Francisco: Jossey-Bass, 1990).

[16]See, for example, J. Richard Hackman, "The Design of Work Teams" in J. W. Lorsch, ed., *Handbook of Organizational Behavior* (Englewood Cliffs, N.J.: Prentice-Hall, 1987).

[17]Nonaka, "Redundant, Overlapping Organization," 51–56.

[18]Mary Lou Davis-Sacks, "Credit Analysis Team," in Hackman, ed., *Groups That Work*, 126.

[19]Hackman, "The Design of Work Teams," 315.

[20]For an overview of information technology support for meetings, see A. Dennis et al., "Information Technology to Support Electronic Meetings," *MIS Quarterly* (December 1988): 591–624.

[21]Susanna Opper and Henry Fersko-Weiss, *Technology for Teams* (New York: Van Nostrand Reinhold, 1992).

[22]Robert Johansen, *Groupware: Computer Support for Business Teams* (New York: Free Press, 1988).

[23]Irene Greif, ed., *Computer-Supported Cooperative Work: A Book of Readings* (San Mateo, Calif.: Morgan Kaufmann Publishers, 1988).

[24]"Workflow: Automating the Business Environment" (Norwell, Mass.: BIS CAP International, 1990).

[25]Cummings and Molloy, *Improving Productivity and the Quality of Work Life.*

[26]Quoted in John Diebold, *The Innovators* (New York: Truman Talley/Plume, 1990): 68.

[27]Zuboff, *In the Age of the Smart Machine.*

[28]Richard E. Walton, *Up and Running: Integrating Information Technology and the Organization* (Boston: Harvard Business School Press, 1989).

[29]Edward H. Rensi, "Computers at McDonald's," in J. Fred McLimore and Laurie Larwood, eds., *Strategies . . . Successes . . . Senior Executives Speak Out* (New York: Harper & Row, 1988): 159–169.

[30]See Tom Richman, "Mrs. Fields' Secret Ingredient," *Inc* (October 1987): 65–72; also see Keri Ostrofsky and James I. Cash, Jr., "Mrs. Fields Cookies," 9-189-056. Boston: Harvard Business School, 1988. The combination of commitment and compliance effects of IT at Mrs. Fields is noted in Walton, *Up and Running.*

[31]Mutual Benefit Life has since undergone financial trauma resulting from its real estate investments and customer psychology about them. Their problems, however, have not involved underwriting processes. The Mutual Benefit Life process innovation is described in James D. Berkley and Robert G. Eccles, "Rethinking the Corporate Workplace: Case Management at Mutual Benefit Life," N9-492-015. Boston: Harvard Business School, 1991.

[32]William Lazonick, "Value Creation on the Shop Floor: Skill, Effort, and Technology in U.S. and Japanese Manufacturing," working paper, Harvard Business School, February 1988.

[33]Robert Hayes and Steven Wheelwright, "Japanese Approaches to Manufacturing Management," in *Restoring Our Competitive Edge* (New York: John Wiley, 1984): 352 374.

[34]Several key issues in workstation-based training are explored in Craig Lambert, "The Electronic Tutor," *Harvard Magazine* (November–December 1990): 42–51.

[35]Personal interviews, Federal Express Corporation, September 1990; also see Amy Bermar, "FedEx Expects Better Service via Networked Training Plan," *PC Week* 5:7 (February 16, 1988): C1–2.

[36]For further detail on this list of motivating characteristics, see J. Richard Hackman and Greg R. Oldham, "Motivation Through the Design of Work: Test of a Theory," *Organizational Behavior and Human Performance* 16 (1976): 250–279.

[37]See, for example, R. Hill, "Working on the Scanlon Plan," *International Management* (1974): 39–43; also E. Smith and G. Gude, "Reevaluation of the Scanlon Plan as a Motivational Technique," *Personnel Journal* 50 (December 1971): 916–923.

[38]See Alan S. Blinder, ed., *Paying for Productivity* (Washington, D.C.: Brookings Institution, 1990).

[39]Reference on Japanese companies in Hayes and Wheelwright, *Restoring Our Competitive Edge: Competing through Manufacturing*.

[40]See, for example, the materials of Organizational Development Resources, based in Atlanta, Ga.

[41]Walton, *Up and Running*.

[42]For more information on this concurrent approach to implementing systems, see Gibson and Davenport, "Systems Change: Managing Organizational and Behavioral Impact," 23–27.

Chapter 6

Creating a Process Vision

Creating a strong and sustained linkage between strategy and the way work is done is an enduring challenge in complex organizations. Because business processes define how work is done, we are dealing with the relationship between strategy and process. Process innovation is meaningful only if it improves a business in ways that are consistent with its strategy. In fact, process innovation is impossible—or at least only accidental—unless the lens of process analysis is focused on a particularly strategic part of the business, with particular strategic objectives in mind. Even the powerful enablers of process innovation discussed above should be applied, not indiscriminately, but in the context of a strategic focus. Earlier, we discussed the selection of strategic processes for innovation. Here, the focus is on embodying an organization's strategy in a vision of the future process state.

Inspiring a vision for operational processes clearly is not the only role for strategy. High-quality, low cycle-time products and services are only useful if they fit the external environment and satisfy a customer demand. This is not always the case with companies that make quality—which is not a strategy, but a way of implementing strategy—a key aspect of their strategies. Examples of this problem abound. A defense industry that has invested heavily in quality at a time when governments are determined to spend less on defense might better have devoted resources to finding commercial markets for its products and expertise. The Cadillac division of General Motors won the Malcolm Baldrige quality award, yet continues to lose market share to Lincoln-Mercury, even among buyers of American luxury cars.[1] This lack of fit between operations and strategy has been called "the six-sigma buggy-whip company" problem.

Congruence or alignment between strategies and processes is essential to radical change in business processes. Strategy and process objectives must reinforce one another and echo similar themes. The concept of strategic "fit" is not new; it underlies the notion of fit as developed by James Brian Quinn,[2] and Anthony Athos and Richard Pascale's "7S" model, which calls for alignment among strategy, structure, style, systems, skills, staff, and superordinate goals.[3]

What is missing in the discussions of strategy experts is the notion of process or process innovation. Fit between strategy and process, always important, is particularly important when the goal is radical business change. A well-defined strategy, in particular, is essential to provide both a context for process innovation and the motivation to undertake it. Radical change cannot be accomplished without clear direction.

But strategy cannot motivate innovation in the absence of a well-defined process vision. A process vision consisting of specific, measurable objectives and attributes of the future process state provides the necessary linkage between strategy and action. Unless such a vision is shared and understood by all the participants in a process innovation initiative—before redesign begins—the effort will all too easily slip from innovation to improvement.

Use of the terms "process simplification" or "process rationalization" in a process change initiative usually indicates that no vision is present, and that only marginal change is likely to be achieved. These terms pertain to the elimination of obvious bottlenecks and inefficiencies, with no particular business vision as context. Rationalization has a long tradition in industrial engineering, as evidenced in a 30-year-old manufacturing-oriented reference note from the Harvard Business School:

> A good manager . . . is expected . . . to show that job improvement or simplification of work is not only important but also is based on common-sense questioning aimed at uncovering the easiest, most economical way of performing a job.[4]

Process innovation is much more than rationalization or simplification, and more than common sense. It questions conventional wisdom about what is easy and economical and thus at times leads to more complex, rather than simpler, processes. Phoenix Mutual Insurance's new underwriting process, which speeded underwriting by making activities parallel rather than serial, is a much more complex process than its predecessor,[5] and Federal Mogul's new process for producing automobile component prototypes, though it has more steps than the old process, takes one-seventh the time. Not only are rationalization and simplification invalid objectives for process innovation, they can be distracting or even misleading as a way of communicating what the effort is all about.

Process visions, like strategies, should be easy to communicate to the organization, nonthreatening to those who must implement (or who are affected by) them, and as inspirational as measurable targets can be. These criteria generally make cost or headcount reduction objectives less than satisfactory, since they are rarely inspiring and are almost certain to be perceived as threatening by employees. When cost reduction is the ultimate objective, it should be mixed with other intermediate objectives, such as time reduction or quality improvement, that might lead to cost savings. In the several firms we have observed in which cost savings is the stated rationale for process innovation initiatives, there has been a notable reluctance to participate in or otherwise aid the program.

Although the empirical evidence is slight, we believe that firms that have quality, improved work life or learning, or time-oriented objectives often achieve cost reduction in their process innovation initiatives, while those that strive only for cost reduction may not achieve even that. The futility of focusing exclusively on cost was observed in an Ernst & Young survey of factors leading to product competitiveness,[6] though it did not take an explicit process perspective.

Process change without strategy and vision seldom goes beyond streamlining, with a resulting incremental reduction in time and cost. Even process improvement is most valuable in a strategic context; streamlining is most useful in areas that really matter to a business. Change is much more likely to be successfully implemented by and to benefit a business when it is focused on what matters most. General Electric's corporatewide "Work-Out" program for eliminating unnecessary work, for example, is focused on no particular business strategy other than productivity improvement and lacks specific objectives regarding the level of work to be eliminated. A masterpiece of employee communication that encourages overworked employees to try to find ways to reduce their own workloads, the program has been praised in the popular press, and analysis has revealed examples of functional productivity improvements.[7] But several managers we interviewed reported that Work-Out had yielded at best incremental results, and that the program's methods were better described as process improvement than innovation.

Far more impressive results have been achieved by General

Electric in a more focused effort at its Salisbury, Maryland, electrical equipment plant. This plant's business strategy was to reverse several years of poor financial results in the circuit breaker business through innovation in process, organizational structure, and technology. The plant adopted self-managing work teams, established electronic linkages with customers, and developed highly visible measures of process performance. These process innovations have reduced customer complaints tenfold, costs by 30%, and delivery cycles from three weeks to three days.

This contrast of two operational improvement initiatives at GE suggests that business units are the appropriate locus for process innovation efforts. When multi-SBU corporations define strategies at the corporate level, they generally involve financial or market share objectives such as GE's desire to be the largest or second-largest business in every industry in which it participates. It may not be possible for an organization the size of General Electric to define at the corporate level strategies that can inspire process innovation in specific businesses.

Key activities required to create visions for process innovation are listed in Figure 6-1. A firm's business strategy provides the overall context for an innovation effort and is assumed to be an input into the innovation initiative. The primary output is process vision, consisting of specific objectives and attributes.

In this chapter, we discuss how business strategy should inform process visions and what role customer perspectives and external benchmarking should play in formulating them, and we consider in depth what their context ought to be, including performance objectives and specific process attributes.

Figure 6-1   Key Activities in Developing Process Visions

- Assess existing business strategy for process directions
- Consult with process customers for performance objectives
- Benchmark for process performance targets and examples of innovation
- Formulate process performance objectives
- Develop specific process attributes

ASSESSING EXISTING BUSINESS STRATEGY

Whereas most recent literature on strategy and vision focuses on the content of strategy and the formulation of strategic position relative to environment, customers, competitors, and so forth,[8] here we are concerned with the implementation of strategy as a means to guide and inspire process innovation. Definitions of strategy and vision abound. We view strategy as long-term directional statements on key aspects of a firm or business unit, and vision as a detailed description of how, and how well, a specific process should work in the future. Vision is thus more tactical than strategy, although both must be formulated at a high level. Moreover, a strategy should be visionary (i.e., look into the future), and a vision strategic (i.e., have a broad, key issue focus). This gives rise to possibilities for confusion.

We assume for purposes of this book that a well-defined business strategy is antecedent to a process innovation initiative. We do not assume that there is a single, correct strategy for an industry or product. Companies can succeed with very different strategies as long as they are well executed.

A defined strategy is a primary determinant in both the selection of, and development of process visions for, processes to be innovated. The roles of business strategy and other aspects of a firm's environment in selecting processes for innovation were discussed earlier; here, we examine the relationships between strategy and process visions.

Because it bears a heavy load in establishing a context for process innovation, a strategy, in addition to positing a broad view of a future state, must also meet the following criteria.

- Strategy should be at least partially nonfinancial. Revenue, profitability, or ROI-based visions often lack meaning for those who will not necessarily benefit from their achievement (i.e., most employees). Nor do such visions help employees understand how their targets are to be achieved. It is, in fact, all too easy for executives to state financial strategies without knowing how they will be achieved. Financial goals, though important, must be combined in a balanced way with process- and product-oriented goals.

- The components of a strategy should ultimately be measurable. Nonfinancial strategies are not as easily measured as financially oriented strategies. Yet a strategy is meaningless unless progress toward its goals can be assessed, a requirement frequently violated in many companies as noted in the literature on nonfinancial performance measures[9] and "strategic control."[10]

- Strategy should focus an organization on specific aspects of its business to which process innovation can be usefully applied. To strategize about becoming a "world-class" company is not terribly useful; becoming the world's best at logistics and distribution is a much more workable strategy.

- Strategy should be distinctive to an industry and company. We are familiar with many steel industry strategies that don't mention steel, airline company strategies that don't mention flying, and so forth. This aspect of strategy, although not important in itself, renders other requirements more likely to be met.

- A strategy should be inspirational. In order to inspire, a strategy must be clearly understood and meaningful to its audience and be the genesis of a considerable amount of work. An inspirational strategy can provide energy when setbacks occur or implementation teams lose momentum. The emotional content of a strategy is sometimes captured in its mission statement. This issue has often been ignored or paid lip service, but companies that have acknowledged it have had successful results.[11]

- A strategy should be for the long term—five or even ten years. It should be broad enough to encompass a variety of opportunistic directions within its overall context.[12] The notion of a long-term yet flexible strategy is captured in the recent literature on "strategic intent."[13] A flexible strategy should be capable of accommodating movement into new businesses built around information and information technology. For example, Baxter Healthcare, American Airlines, and a number of other firms that have created electronic distribution processes now use them to distribute products well outside their mainstream businesses (in Baxter's case, office supplies, in American's, flowers and candy).

- The method employed to create a strategy, like the change

levers that enable process innovation, should be broadly focused and address key tools for change. The planning method
that best seems to do so is scenario-based planning.[14]

Among the companies pursuing process innovation that have
strategies well suited to their process needs is Continental Bank,
which has a strategy that calls for being a business bank that
provides excellent service to a small number of medium-sized
customers. Continental Bank's chairman Thomas Theobald's frequent spoken and written comments about the need for radical
change in banks and banking regulation—e.g., "Frankly, I just
don't think that cuts of 10 or 15 percent will do the job"—motivate
innovation rather than improvement.[15] Continental is struggling
to adhere to its strategy and achieve strong financial results, but
it clearly has come a long way since its bankruptcy in the early
1980s.

Rank Xerox U.K.'s strategic shift from selling copiers to marketing document management systems might seem a subtle distinction, but the firm's executives have developed explicit measures
of success (e.g., 50% of revenues from systems by 1990) and oriented new process designs toward improving these measures.[16]
In contrast, IBM has launched an ambitious process innovation
program based on a strategy of "market-driven quality," which,
though admirable and likely to yield some benefits, may not be
sufficiently focused to bring about truly radical change or inspire
specific and aggressive process visions. Indeed, in attacking 18
key processes simultaneously, the company is more likely to
achieve process improvement than innovation.[17]

Satisfying the foregoing criteria for strategy improves a firm's
chances of success in the search for process innovation. Yet any
of these requirements can be taken too far. Nonfinancial strategies make sense only to the degree that they lead to better financial performance. Overemphasis on measurability and an
overly detailed focus can inhibit both a strategy's long-term ap-
plicability and its ability to inspire employees. Strategy, moreover, provides only an internal perspective in creating a process
vision. There are many ways to implement even a well-defined
strategy. Many innovative ideas for process design and the use
of enablers can come from sources outside the organization, including customers, competitors, and firms in completely different

businesses. To incorporate these sources into its visioning activity, an organization must include in its process innovation initiative both a structured exercise to understand customer perspectives and a round of external benchmarking.

CUSTOMER INPUTS INTO PROCESS VISIONS

A key aspect of creating a process vision is to understand the customer's perspective on the process. As noted previously, the customer of a process can be either internal or external to the firm; in practice, most firms are more concerned about the perspectives of external customers, and therefore place a high priority on customer-facing processes.

Asking customers what they require of processes serves multiple purposes. In the context of creating visions, the customer perspective furnishes both ideas and objectives for process performance. Seeking customer input also demonstrates a desire for a close relationship, although this input must be actually factored into process designs to fully achieve this objective. Finally, new processes may require that customers change their own behavior for the process to be fully effective. Seeking input at an early stage starts building customer commitment to the needed changes and lets customers begin their own process transition.

Of course, most processes have multiple customers. It is often not very useful to treat all customers equally; a written questionnaire sent to hundreds of customers, for example, will seldom deliver meaningful results. It is much more valuable to select a limited set of customers, based either on their importance to the firm or their ability (and willingness) to furnish thoughtful input, and spend more time with each one. Focusing on a few customers also allows for less structured interview processes in which innovative ideas can be sought.

The type of inputs that should be solicited from customers are broad, encompassing desired process outputs, performance, flow, enablers, and other relevant factors. The method of customer contact may take several forms. A focus group may be the best way to deal with individual, rather than organizational, customers. Some firms prefer to send consultants to talk to customers, so that they will not hold back on their true feelings. Other firms would never pass up an opportunity to build closer personal relationships with customer executives. Sometimes the solicitation of opinions from customers can be done in an informal setting. For example, the discussions that led to Procter & Gamble's continuous replenishment processes for Wal-Mart took place during a fishing trip by Sam Walton and a P&G executive.

In our experience, customers rarely provide breakthrough ideas for process innovations. Instead, their objectives are to improve the existing process incrementally: "I would like to have more on-time deliveries." These types of inputs are important, however, because they specify the areas in which innovation should take place. If the customer feedback process is somewhat iterative, taking place throughout a process innovation initiative, customers can react to successively more concrete process designs. As is true for user participation in the design of information systems, customers often do not know what they want until they see what they can get, or until they see something that they know they don't want.

PROCESS BENCHMARKING

As practiced in the quality movement, benchmarking helps companies formulate objectives for continuous improvement programs. But it can also be an effective tool for determining process objectives and identifying innovative process attributes. Insofar as it enables companies to look outside for alternative ways of designing processes, benchmarking can help to break a company's inwardly focused mind-set.

Most appropriate for the purposes of process innovation is a "best practice" or "innovation" benchmark that selects companies on the basis of the performance of a particular process, without regard to the industry, and addresses specific innovations and uses of change enablers as well as overall process performance. A company attempting to innovate its order management process, for example, might study Digital Equipment Corporation's expert systems for automated configuration or USAA's empowered customer service representatives, whereas a firm studying new product development might analyze how J.C. Penney employs videoconferencing to place new fashion designs in stores. Benchmarking can identify realistic performance objectives and target characteristics for companies to match or surpass, information that can be used during innovation brainstorming workshops to fuel the redesign process.

Innovation benchmarking need not focus on the traditional concerns of benchmarking. Process measures and costs may help to establish objectives for a new process, but even a poorly performed process in a poorly performing company can have innovative aspects. It is also important to benchmark distinctive uses of enablers and innovative work designs. Even relatively narrow aspects of processes can be worthy of analysis. Several firms that are redesigning their order management processes, for example, have visited a division of AT&T in which field personnel use laptop computers and portable networking to work without offices. This innovation, although it does not comprise a complete process, can be an important component of an order management approach.

Other industry innovations should not be neglected in formulating the objectives and attributes of a process vision, but managers may be more readily disposed to adopt innovations for which there is evidence of use in their own industry. Benchmarkers must therefore strike a balance in presenting valuable cross-industry innovations and clearly competitive, relevant-within-industry innovations likely to gain quick acceptance.

The sources of benchmarks are varied, ranging from company visits to telephone discussions with consultants and executives in other firms to industry publications and academic case studies. Because third-party accounts of process innovations may gloss over important issues or stop short of the final chapters of a story, a company is wise to contact benchmarked organizations directly at some point in the benchmarking process. Details of benchmarking, such as the etiquette of contacting firms, are similar for quality and innovation-oriented benchmarking (see Camp[18] for specifics).

LINKING STRATEGY AND EXTERNAL INFORMATION TO PRODUCE PROCESS VISIONS

Strategy, customer perspectives, and external benchmarks are necessary, but not sufficient, to establish the context for process innovation. For a process to be transformed, the context must be made explicit and operational through a set of visions that define the desired process functionality, specify change objectives for the redesign of the process, and identify qualitative attributes of the process's future state. These visions provide necessary direction for the design team.

Process visions link strategy and action; they translate high-level strategies into measurable targets for process performance and understandable characteristics of process operations, and they set targets both for the designers of a process and for those who must subsequently manage it. Just as analysis of corporate strategy is combined with information from external sources to create process visions, so process visions give rise to objectives and attributes (see Figure 6-2).

FORMULATING PROCESS OBJECTIVES

Process objectives include the overall process goal, specific type of improvement desired, and numeric target for the innovation, as well as the time frame in which the objectives are to be accomplished. Both general process functionality and change goals should be addressed by these objectives.

Process objective creation begins with a vision team asking itself, and key stakeholders, "What business objective is this process supposed to accomplish?" This analysis should broadly address

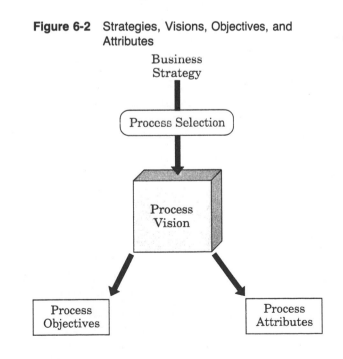

Figure 6-2   Strategies, Visions, Objectives, and Attributes

Business
Strategy

Process Selection

Process
Vision

Process
Objectives

Process
Attributes

the functions and value the process is expected to bring customers. To avoid what Ted Levitt has called "marketing myopia," it should concern itself not only with traditional outputs of the process, but also with real customer needs. A team redesigning a company's order management process might discover, for example, that its customers need additional services to take better advantage of the company's products, or customers might be interested in purchasing products in a less-finished state. The analysis should also address the aspects of the business that are most important to customers. Firms sometimes find that they are focusing on one process objective when another is more important to customers— for example, stressing product variety when customers want faster, more reliable delivery, or striving for cost reduction when customers would willingly pay more for a better product or service.

Process objectives must be quantified as specific targets for change. Examples of quantitative process objectives for various industries might include:

- reduce new drug-development cycle time by 50% in three years;
- double customer service satisfaction levels in two years;
- reduce involuntary employee turnover to 10% by the end of the next fiscal year; and
- reduce processing costs for customer orders by 60% over three years.

Process objectives should be derived from strategy. Continental Bank's strategy of being a business bank led naturally to a process objective of doubling relationship managers' time spent with customers. Although this objective was appropriate, the executive who "owned" customer contact (now CEO of another bank) chose not to publicize it to the relationship managers.

Rank Xerox U.K.'s process vision was not so easily derived from its strategy of selling systems rather than copiers. The strategy applied to numerous different processes and was difficult to translate into specific redesign goals for any one. The substantial benefits realized from the first round of process redesign (e.g., reducing jobs not involved in customer contact and reducing order delivery time from 33 to 6 days), did not reflect the level of business change desired by the senior management team. Consequently, the company is undertaking a second round of process innovation in which the objectives are more specific and more closely tied to current strategy. This round should be much easier than the first, because the company now has a strong process orientation.

Process objectives, like strategies, should meet a number of established criteria. For example, the level of change targeted should be radical. This usually means greater than 50% improvement. Some firms have much more ambitious goals. IBM, for example, is attempting to reduce time, cost, and defects by a factor of 10 by 1992 and by a factor of 100 by 1995. However unlikely the achievement of these goals for most processes, establishing them has clearly stimulated a great deal of work by process design teams, and some improvement is certain to be achieved.

DEVELOPING PROCESS ATTRIBUTES

Process attributes, the descriptive, nonquantitative adjunct to process objectives, constitute a vision of process operation in a future state. They address both high-level process characteristics and specific enablers. Process attributes for an order management process, for example, might specify that the process will employ expert systems-based credit checking, automated proposal generation, increased worker empowerment, and a financial structure resembling dealerships for customer-facing teams. Other attributes might specify that the process be performed by one person or team, and that credit, shipping, and scheduling functions will be performed by the customer-facing individual. It is sometimes useful to categorize attributes as "technology," "people," "process outputs," and so forth.

Process attributes might be considered principles of process operation.[19] Like principles, they are simple statements that describe an organization's philosophy and intent regarding process operations and can be an effective means of engaging senior managers in discussions about visions for new processes.

The enablers of process innovation that have been identified as relevant in earlier stages of the innovation initiative also become attributes of the process. These may involve information, information technology, or organizational and human factors.

An example of an organizational attribute is to collapse the division of labor in a process, that is, to organize it in such a way that the entire process is overseen by a single employee or group.

In many banks, for example, relationship managers are responsible for handling all aspects of a customer relationship, and many hospitals employ case managers to coordinate all aspects of a patient's stay.

An example of a human factors attribute is to push decision making down in the organization. One manufacturer that had made pricing decisions the responsibility of the division general manager moved that decision making down in the organization to cross-functional product development teams; in doing so, it realized overall cycle-time improvements. Other companies, by applying this attribute, have eliminated layers of management and hierarchy.

A common technology-oriented attribute is the offloading, whenever possible, of process activities to process customers by giving them access to the provider's computer systems. One future vision of Federal Mogul's customer-driven product development is to allow customers to examine its on-line product catalog to determine whether an existing product might satisfy their requirements, an activity currently performed by product engineers. Federal Express and other firms in the shipping industry have given customers terminals to check the progress of parcel shipments.

These are only a few examples of attributes that might be employed within process visions. The enablers selected by a design team, along with inputs from customers, external benchmarking partners, and the business strategy, should suggest many more.

When NCR reengineered its order management process in 1988, a set of attributes (called "goal states" by NCR) were developed as part of the visioning activity. Managers of the initiative credited the goal states with presenting "a case for action" and being a vehicle for achieving buy-in by process owners and participants. Among the 20 attributes (some were really performance objectives) they declared were the following:

- link order management systems worldwide, but keep them local;

- create automated sales assistant tools for product information and pricing;

- offer direct shipments from manufacturing to customers—no warehouses; and

- render invoices and payments electronically.

Since the original attributes were defined at NCR, they have continued to evolve. As the envisioned process was implemented, both the business and relevant technology continued to change; it was only natural that the process attributes would also change through the implementation period.

THE VISIONING PROCESS

Process objectives and process attributes are derived from multiple sources—among them, analyses of corporate strategy and vision, high-level overviews of the roles of technology and people (as both opportunities and constraints), customer interviews, benchmarking of the best processes in other companies, and a firm's performance objectives—during visioning sessions at the beginning of a specific process initiative.

These visioning sessions should be a series of workshops, with increasing specificity about the vision at each step (see Figure 6-3). In earlier stages of the process, the focus is on attributes and objectives, as described above. After these have been articulated, later workshops can address the critical factors for the successful implementation of the vision, and any barriers that might stand in the way.

Regardless of the topic focus, all aspects of the vision must be stated with a high degree of specificity. The specificity of process visions is the source of both their power and their difficulty. Specific attributes and measurable objectives are powerful because they clearly express the purpose of process innovation efforts. But because they may be formulated well before the process team begins detailed analysis of the existing process, they are difficult to formulate with accuracy, and they rely on prediction more than precision. A process vision should therefore be determined on the basis of what is necessary from a business standpoint, rather than what seems reasonable or accomplishable.

Process objectives should be stretch targets for an organization. Unless their reach seems to exceed an organization's grasp, they will not motivate design teams to go beyond incremental improvement. An improvement objective of 10% permits a team to streamline a process—to eliminate a step here, a person there.

Figure 6-3   The Visioning Process

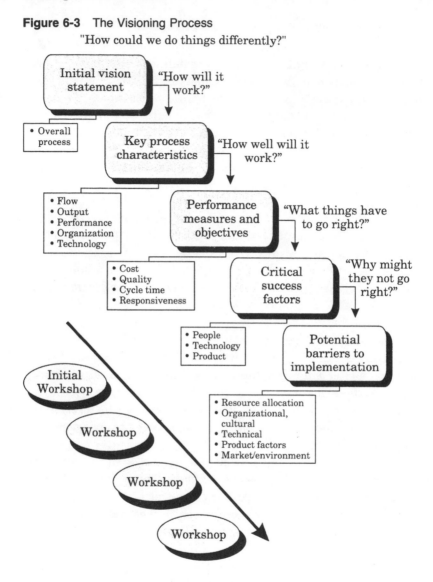

Specifying an objective of 100% improvement ensures that a process will be examined with an eye toward radical change. The purpose of stretch targets is primarily behavioral—to let designers and implementers know that innovation and creativity are expected. To target low levels of improvement, or to say, "Let's just look at the process and see what we can get," is to invite marginal improvement at best.

A vision must not be completely unrealizable. A design team that perceives, for example, that a stretch objective cannot be achieved—either because it is unreasonable or because environmental constraints are too great—becomes disillusioned, a lesson learned in research on high-commitment organizations.[20]

Process objectives and attributes are important not only for new, but also for existing processes. As explained in the next chapter, it is important to understand an existing process before designing a new one. This understanding must include measurement of the aspects of the process upon which the new process is to improve, and comparisons of the attributes of the old and new processes. Without an explicit process vision for a new process, it is difficult to know what to examine in the old one. "Safe" measures, such as the cost of and time consumed by an old process, should not be examined to the neglect of other, more vision-specific measures.

Process visions are a means of enabling some degree of wider participation in an activity that needs to be driven from the top of the organization. They place on the shoulders of lower-level members of process teams the work of designing new processes to achieve objectives and attributes clearly stated by senior management. Because process team members are typically drawn from corporate staff and middle management groups, it is not accurate to refer to process implementation as "top-down vision, bottom-up implementation"; at best, implementation is "middle up." The middle ranks of an organization will also be responsible for managing the new process to achieve full benefits.

Process visions and the strategies that support them are key elements of a context for process innovation. Like other aspects of management practice, clear strategies and visions are a good idea for all companies, but they are essential to process innovation.

Senior management plays an important role in creating process visions. The foundation of Hallmark's innovation efforts is a vision of the company's ideal future state as articulated by its top 40 managers.[21] Process innovation teams at Xerox are guided by Xerox 2000, a corporate, high-level description of core competencies for the millennium developed by senior management, with strategic goals for specific process initiatives supplied by divisional management. Some senior managers must be participants in process visioning activities; others must be aware of the process visions that result.

SUMMARY

The creation of process visions relies on assessing an organization's strategy, gathering external inputs into process design and performance, and translating this information into specific process objectives and attributes. The relationships among these factors must not be one-way. Lessons learned through process innovation should influence strategy. If process orientation is to become a key determinant of how business is conducted, results of process management should feed back into all aspects of the organization. Strategies should eventually be phrased in terms of which processes must be changed, and their success measured through process-generated information. Objective setting should be tied to results achieved in previous processes.

As firms become more comfortable and experienced with information and information technology as enablers of process innovation, strategies and objectives should shift to better exploit these competitive resources. Information management, when it becomes a core competence, opens entirely new competitive alternatives, including the sale of information, information management services, and alternative products through information-based channels.[22]

Thus far, we have focused primarily on shaping designs for new ways of doing business. But the current process must also be the focus of analysis, both in terms of understanding and measurement and in terms of short-term improvement. It is to this that we turn in the next chapter.

Notes

[1]Jeremy Main, "Is the Baldrige Overblown?," *Fortune* (July 1, 1991): 62–65.

[2]James Brian Quinn, *Strategies for Change: Logical Incrementalism* (New York: Richard D. Irwin, 1980).

[3]Anthony G. Athos and Richard T. Pascale, *The Art of Japanese Management: Applications for American Executives* (New York: Warner Books, 1982).

[4]"Reference Note on Work Simplification," 9-609-060. Boston: Harvard Business School, 1961.

[5]B. Birchard, "Remaking White Collar Work," *Enterprise* (Digital Equipment Corporation: Winter 1990): 19–24.

[6]William Smith, *The American Competitiveness Study: Characteristics of Success* (New York: Ernst & Young, 1990).

[7]Thomas A. Stewart, "GE Keeps Those Ideas Coming," *Fortune* (August 12, 1991): 41–49.

[8]Michael E. Porter, *Competitive Strategy: Techniques for Analyzing Industries and Competitors* (New York: Free Press, 1980) and Michael E. Porter, *Competitive Advantage: Creating and Sustaining Superior Performance* (New York: Free Press, 1985).

[9]Robert G. Eccles, "The Performance Measurement Manifesto," *Harvard Business Review* (January–February 1991): 131–137.

[10]See, for example, Michael Goold, "Strategic Control in the Decentralized Firm," *Sloan Management Review* (Winter 1991): 69–81.

[11]Andrew Campbell, Marion Devine, and David Young, *A Sense of Mission* (London: Economist Books, 1990).

[12]See Daniel J. Isenberg, "The Tactics of Strategic Opportunism," *Harvard Business Review* (March–April 1987): 92–97.

[13]Gary Hamel and C. K. Prahalad, "Strategic Intent," *Harvard Business Review* (May–June 1989): 63–76.

[14]For a discussion of scenario-based planning, see Peter Schwartz, *The Art of the Long View* (New York: Doubleday Currency, 1991).

[15]Thomas Theobald, "Restructuring the U.S. Banking System for Global Competition," *Journal of Applied Corporate Finance* 3:4 (Winter 1991): 21–24.

[16]These Rank Xerox strategies have been supplanted by others since their original creation. See "Rank Xerox U.K., Office Systems Strategy (C): Developing the Systems Strategy," case study from Henley—The Management College, September 1988. Also see Thomas H. Davenport and James E. Short, "The New Industrial Engineering: Information Technology and Business Process Redesign," *Sloan Management Review* (Summer 1990): 11–27.

[17]Though innovation initiatives exist for these 18 processes, there are some signs that IBM is beginning to focus its major initiatives on a smaller number of processes.

[18]Robert C. Camp, *Benchmarking: The Search for Industry Best Practices That Lead to Superior Performance* (Milwaukee, Wis.: ASQC Quality Press, 1989).

[19]Thomas H. Davenport, Michael Hammer, and Tauno J. Metsisto, "How Executives Can Shape Their Company's Information Systems," *Harvard Business Review* (March–April 1989): 130–134.

[20]Richard E. Walton, "From Control to Commitment: Transformation of Workforce Management Strategies in the United States," in Kim B. Clark, Robert H. Hayes, and C. Lorenz, eds., *The Uneasy Alliance: Managing the Productivity-Technology Dilemma* (Boston: Harvard Business School Press, 1985).

[21]Allan E. Alter, "The Corporate Make-Over," *CIO* (December 1990): 32–42.

[22]The concept of core competence is defined and discussed with reference to strategy in Gary Hamel and C. K. Prahalad, "The Core Competence of the Corporation," *Harvard Business Review* (May–June 1990): 79–91. Some Japanese firms appear to have this core competence; see James Matarazzo and Laurence Prusak, *Japanese Success and Information Management* (Washington, D.C.: Special Libraries Association, 1991).

Chapter 7

Understanding and Improving Existing Processes

It is important to understand an existing process before designing a new one. Some approaches to process redesign and reengineering do not include this step, and some companies have omitted it in their process innovation initiatives—to their regret. One process team at a large telecommunications firm, for example, estimates that not taking the time to understand an existing process left it unable to establish the benefit of adopting the new process. The team found it difficult to get funding for the initiative, resulting in an 18-month delay.

There are at least four reasons to document existing processes before proceeding with innovation. One, understanding existing processes facilitates communication among participants in the innovation initiative. Models and documentation of current processes enable those involved in the innovation activities to develop a common understanding of the existing state. This aspect is particularly important in the case of business and management processes performed by professionals or other employees engaged in relatively unstructured, nonroutine work, which such individuals find difficult to view as a process.

For example, a large defense contractor attempting to innovate its information-systems development processes was unable to get its professionals to recognize (or admit) that a process underlay the work they performed. The firm was forced to use an incremental improvement approach to the development activities, with no process context or measurement of either current or improved activities. The merchant banking group of a major investment bank wanted to take a process perspective to define more systematic ways of identifying new business opportunities after investment opportunities took a downward turn in the late 1980s.

137

The group ultimately failed because the senior bankers were unwilling to shed their self-image as dealmakers and view their work in process terms.

Two, in most complex organizations there is no way to migrate to a new process without understanding the current one. Current process documentation is an essential input to migration and implementation planning, useful for understanding the magnitude of anticipated change and the tasks required to move from the current to a new process.

Three, recognizing problems in an existing process can help ensure that they are not repeated in the new process. IBM Credit Corporation's first redesign of its financing-quote preparation process duplicated the sequential queues of the previous version. When the redesign results did not meet expectations, an examination of the new process design attributed unsatisfactory cycle time to the queue structure. A second redesign initiative that combined the multiple functions of quote preparation into one job achieved tenfold reductions in cycle time and tenfold increases in the number of quotes prepared.

It is not unusual for process problems to go unrecognized until an entire process is scrutinized. Process actions in one department may have a negative impact on another organizational unit that goes unheeded until the departments are brought together in a process innovation study. Consider a manufacturing firm that attempted to shorten the product development cycle by modifying incentives in the engineering department. Product engineers, compensated on the basis of throughput, began to skip the manufacturing review step in order to meet their targets. This omission soon resulted in a noticeable increase in the number of designs that could not be manufactured. Failure to consider overall process implications rendered what would otherwise have been a good decision based on departmental objectives a bad decision.

Finally, an understanding of the current process provides a measure of the value of the proposed innovation. Baseline data collection is one facet of current situation analysis. Given a process objective of reducing cycle time, for example, baseline data collection would need to include measurement of elapsed time for the current process.

EXISTING PROCESS ACTIVITIES

Several books have been written about understanding and improving existing processes.[1] Our discussion here is limited to current process analysis in the larger context of an innovation initiative. In general, improvement initiatives require a great deal more detailed information on current processes than innovation initiatives. Analysis of the existing process in an innovation context should consume a number of weeks, not months.

Key activities in existing process analysis and improvement are listed in Figure 7-1. The first is documenting current process flow. Often existing processes have never been described or even viewed as processes. Describing a process is central to the purpose of process communication discussed earlier. Process participants who do not view their current activities in process terms are not likely to readily adopt a revolutionary process.

Even when the existing approach to work has already been viewed in process terms, it may not have been analyzed in sufficient breadth. Process analysis in quality initiatives usually deals with narrow processes, which often vary from one part of a company to another. For example, when analyzing the existing process for order management at a manufacturing firm known for its quality programs, we visited three billing centers with identical responsibilities. Each center had identified its processes—but one had 20, one 50, and the third 300—all for an activity that was only a part of a single order-management process in the innovation initiative. Making these divergent process analyses comparable, and "rolling them up" to yield composite

Figure 7-1   Key Activities in Understanding and Improving Existing Processes

> • Describe the current process flow
>
> • Measure the process in terms of the new process objectives
>
> • Assess the process in terms of the new process attributes
>
> • Identify problems with or shortcomings of the process
>
> • Identify short-term improvements in the process
>
> • Assess current information technology and organization

measures for the cost and time of managing orders, were perhaps more difficult than if the company had never viewed its work in process terms.

Because it will be used as a baseline for comparisons with the new process, the existing process should be assessed in terms of the same criteria employed for the new design. Thus, the scope of the existing process should be the same as that envisioned for the new process. Furthermore, the current process should be measured on the specific performance objectives, and assessed for the relative levels of process attributes, identified in the process vision. This narrowing of the assessment task helps reduce the time and effort required for the current process analysis.

If, for example, the visioning activities for a new process yield as objectives cycle-time reduction and output quality improvement, these criteria should be the focus of current process measurement activities. If the process attributes in the new vision involve a single level of approval for customer transactions, the number of approvals required in the current process should be known.

Current process analysis should also include assessments of information technology and organization. Assessment of the existing information technology architecture should include existing applications, databases, technologies, and standards. Assessment of the organization should include job descriptions, skills inventory, and knowledge of recent organizational changes (e.g., implementation of a new performance measurement system).

IMPROVING THE CURRENT PROCESS

To say that process innovation is more comprehensive and ambitious than rationalization and simplification is not to suggest that opportunities for the latter types of improvement be ignored. Improving existing processes is a natural follow-on to documenting them. Analysis provides employees with an opportunity to document problems they may have known about for years, and, at the same time, because most companies do not study processes on a regular basis, examining a process as a whole often highlights long-standing problems, such as bottlenecks, redundancies, and unnecessary activities, that have gone unrecognized.

A number of companies engaged in process innovation initiatives are coupling short-term improvement and breakthrough innovation. IBM expects its process-modeling efforts to yield opportunities for short-term improvement, namely, elimination of redundant activities and defects, cycle time reduction, and cost savings. These short-term improvement efforts are being undertaken largely by the employees who perform the processes day to day. Xerox is using a two-team approach in which process innovation and short-term improvements are identified in parallel. Membership on the current state team is different from that of the future state team, with limited interaction beyond the transmission of current results. Faxon, which is taking a similar approach, has documented problems identified during the course of data gathering and developed an implementation plan to rectify these situations while systems development work proceeds.

A communication mechanism is needed to enable innovation teams and improvement groups to share important information about current process and future visions. Some firms practicing both innovation and short-term improvement believe that only improvements consistent with the future vision of a process should be implemented. At Prudential Insurance, an initiative to create a process vision is under way, although the vision may not be implemented in an innovation initiative. Rather, the vision may be used only to determine whether short-term improvement activities are headed in the right long-term direction.

Unless they are clearly distinguished, undertaking innovation and improvement activities concurrently can be confusing, but in a very large organization, it may be the only way to achieve short-term benefit. Inasmuch as process innovation typically takes several years to implement fully, short-term improvements offer a way to begin to deliver results. Short-term improvements can be implemented while the innovation work proceeds. Some firms are viewing process improvement in the near term as a vehicle for funding process innovation over the long term.

TRADITIONAL APPROACHES TO PROCESS IMPROVEMENT

Approaches for improving business processes are both abundant and diverse. Information systems, industrial engineering, operations research, and management accounting are among the disciplines represented. We consider the approaches presented in Figure 7-2, namely, process value analysis, activity-based costing,

Figure 7-2   Process Methods Overview

Approach	Objective	Tools / Method	Roots
Activity-Based Costing	Cut cost	Cost buildup over process/value-added analysis	Accounting for product line selection
Process Value Analysis	Streamline a single process/reduce cost and time	Value analysis for each process step	Consulting approaches
Business Process Improvement	Continuously improve one or all processes in terms of cost, time, and quality	Process step classification, quality tools	Total quality management
Information Engineering	Build a system along process lines	Descriptions of current and future processes	Systems analysis
Business Process Innovation	Use change levers to radically improve key processes	Change levers, future vision	Competitive systems

business process improvement, information engineering, and business process innovation.

An overview of these approaches serves two purposes. First, in explaining why none is capable of achieving radical improvement, it helps firms understand the differences between traditional approaches and process innovation. Second, it identifies tools and techniques that are applicable and useful in the process improvement phase of a process innovation initiative. Each approach is most relevant to a particular process environment. A firm can choose among these improvement methods, or apply particular frameworks and techniques to create a novel approach.

Activity-Based Costing

Management accounting is supposed to provide information needed for executive decision making regarding product mix, profitability analysis, and so forth. The core of traditional management accounting, expense allocation based on the direct labor content of a particular activity, is becoming increasingly meaningless as the direct labor content of most processes is reduced or even eliminated.[2] In the competitive environment of the 1990s, success and profitability rely on more than cost containment. Today's management accounting systems must take account of such nonfinancial factors as quality, flexibility, and time to market.

The objective of the activity-based costing approach to management accounting is to determine the resources required to produce a particular product or serve a particular set of customers. The approach implies a process perspective, since it is impossible to understand the resource needs of a product or set of customers without examining the production process. Many companies have discovered that they can use the accurate product-cost information provided by activity-based costing not only to determine which products or customers are profitable or unprofitable, but also to improve a given process.[3] Improvement opportunities in the context of activity-based costing arise in two ways: (1) the process includes analysis of cost drivers and nonvalue-adding activities, and (2) the information produced can be used by employees and management to measure continuous improvement, particularly when the primary objective of an innovation initiative is cost reduction.

Few companies that have used activity-based costing in a process management context have achieved radical process improvements; incremental improvement is more common. Level of improvement may be a function of project scope. Development of an activity-based costing system requires analysis down to the lowest level of activity within a company. Changes at this level, unless they lead to broad product-line restructuring, are likely to produce incremental change at best. Robin Cooper and Robert Kaplan point out that it is impossible to control expenses at the macro level of the company,[4] yet it is there that the opportunity for innovation is greatest.

Cost management systems are a major component of performance management in most corporations, and process performance measurement is a requirement for the ongoing management of an innovated process. Activity-based costing may provide a model for process performance-measurement systems that need to be developed for new processes. For example, activity-based costing systems include nonfinancial performance measures and feedback for continuous improvement, which are important aspects of innovation performance measurement.[5]

Understanding existing processes in activity-based costing involves two steps. In the first, the activities to be performed are identified. Some sort of process model, such as a work flow diagram, is typically used to document process structure. The second step involves attributing costs to activities based upon consumption of resources. This step includes analysis of cost drivers such as machine setup. Outputs of the cost attribution step include a cost buildup diagram that shows the cost and time associated with each activity over the entire process.

Opportunities for improvement arise out of detailed analysis of current process operations, and problems are documented during the course of understanding process activities. It is this level of scrutiny that gives rise to opportunities for streamlining and rationalization.

Process Value Analysis

Process value analysis, or activity value analysis, as it is sometimes called, was developed and popularized by management consultants during the 1970s and 1980s (although some firms have been performing these analyses since the 1950s) as a way to reduce overhead costs. This approach, which offers a systematic way to analyze the costs and value associated with various processes, was intended as a bottom-up response to top-down cost reduction programs, many of which involved indiscriminate percentage-based reductions of cost or employees across an organization, resulting in the loss of valuable employees and programs.

Process value analysis is a fairly straightforward approach that involves studying process components and activities to understand process flow. Taking the current process as a point of departure, it documents elapsed time and expense for each activity. Requirements of customers, both internal and external, are solicited and used to test the value of each activity. Activities that add no value to a process output (in the eyes of the customer) become candidates for elimination.

The primary model used in process value analysis is a cost buildup chart, similar to that used in activity-based costing. It depicts the sequence of process activities and the time and cost associated with each activity. The value test is then applied to each activity to drive both time and cost out of the process.

The main limitation of process value analysis is that it is a one-time solution to a problem. Because it provides no mechanism for sustaining improvements, companies that employ process value analysis frequently revert to old practices within a year or two. This is true of process innovation as well, of course, and constitutes a strong argument for establishing continuous cycles of improvement and innovation that yield sustained benefit over time.

Practitioners of process value analysis have recognized the value of applying value analysis methods at higher levels in the organization through executive participation.[6] However, there is no mention or consideration of vision creation or change enablers that can introduce innovation or more dramatic change into the overall process.

Quality-Based Improvement

James Harrington defines an approach to business process improvement with foundations in total quality management.[7] This systematic, step-by-step approach to streamlining processes and establishing a culture of continuous improvement is robust and comprehensive, but certain fundamental characteristics make it unlikely to achieve breakthrough results. Nonetheless, business process improvement remains a valid way to achieve incremental process improvements on a broad scale.

Business process improvement, like process innovation, maintains an enterprisewide scope, establishes improvement objectives based upon business goals, and provides mechanisms for linking multiple improvement initiatives. But, unlike process innovation, it rejects the notion of using change enablers during improvement activities. Harrington states emphatically that information technology, which he equates with automation, must be considered only after a process has been improved. Moreover, the streamlining approach that he describes is a step-by-step method that employs 13 techniques applied sequentially to the process in its current state to yield improvements of different types. A systematic approach that never completely challenges the initial process structure presents little possibility of creative breakthrough or radical change.

Nevertheless, a number of the tools and techniques described by Harrington can be used on the current process in the context of an innovation initiative. Process diagramming, or flowcharting, for example, can be used to understand existing processes. The flow chart, which Harrington defines as "a graphical representation of process components and interfaces," has value primarily as a communication mechanism that can be used throughout the approach to document the evolution of process work flow as streamlining techniques are applied.

Different types of flow charts are employed in business process improvement. Process documentation and analysis are viewed as iterative processes. High-level block diagrams used to develop a preliminary understanding of the process and its boundaries finalize the scope of improvement activities. Thereafter, detailed flow charts are developed that will serve as the foundation of improvement activities.

Harrington also provides useful perspectives on the formation of executive and operational teams for process work, and on the establishment of a continuous process improvement culture. Although his approaches require high levels of commitment and discipline on the part of adopting firms, they are necessary to the creation and maintenance of a culture of continuous improvement.

Informational Systems

Companies have applied a number of different approaches to process innovation. Although operational improvement per se has not been a traditional objective of systems development, many systems development methods imply such a goal.[8] Several current systems-development approaches document both current and future process states, the future state embodying systems requirements that will improve the current operation in some way. It is also our experience that systems developers frequently argue that they have always been improving and innovating business processes.

Information engineering, a rigorous approach to planning, analyzing, designing, and constructing information systems, became one of the most widely employed systems development approaches during the 1980s and early 1990s.[9] A strength of information engineering as a systems development approach is its data orientation. Whereas traditional systems-development methods have largely focused on the processes and calculations a system needs to perform, information engineering presents a method for considering data as a separate and independent entity.

Information engineering, like process innovation, takes a macro-level perspective. It is a top-down approach that translates strategic business objectives into information systems. Analysis of data and processes begins with the development of enterprise-wide models, representations of the entire organization or major business unit, which are subsequently decomposed in a series of transformations. The resulting models document detailed systems requirements, such as action diagrams and database design specifications. Information engineering has the organizationwide perspective needed for process innovation.

A number of companies, including Xerox and IBM, have attempted to use information engineering to redesign key processes. Although the approach has value for some aspects of process innovation (documenting process designs, for example), a number of firms have ultimately been unsuccessful in using information engineering methods to achieve substantial business change. But these organizations have learned from their early efforts; although they no longer use information engineering to effect process change, they do employ it to build systems to support new process designs.

Among the characteristics of information engineering that make innovation and radical change difficult to achieve is the subordination of process to function. A process can exist only within a single business function such as marketing or finance. Information engineering does not accommodate business processes that cross (as most do) organizational boundaries. To innovate using information engineering would entail creating a broader set of process entities, which would require substantial modifications to the methodology.

Moreover, most information engineering-based efforts are not charged with innovating business processes. The usual goal is to develop a new information system, and systems development is complex enough without expanding the scope to innovate the business process.

Finally, once a project moves beyond planning and is partitioned into business areas, the enterprisewide context is lost. Process innovation requires that the interfaces between subprocesses be managed actively and reinnovated if needed. Information engineering provides no mechanism for doing so. Once a business area is partitioned, analysis, design, and construction proceed independently without regard for what is occurring in other business areas.

The information engineering-based approach that IBM is using to innovate key business processes on a corporatewide basis as part of its attempt to "use information strategically" combines process and data modeling to document current process structure.[10] IBM believes that it needs the process and data models information engineering produces to identify the interfaces where business processes cross organizational boundaries, but we believe that this approach is more likely to yield process improvement than process innovation.

This is not to suggest that no company has successfully developed an innovative process design using information engineering. Faxon, a periodicals clearinghouse, used an information engineering-based systems-development methodology to determine information systems requirements for managing orders. Serendipitously, Faxon was also pursuing quality management efforts and had completed a thorough analysis of the work flow for the same process. Analysis from the two efforts is currently being used to identify opportunities for process innovation, taking advantage of capabilities the new information system will offer.

The information systems methodologies most useful to process innovation are process and data modeling. Process modeling, which can take a variety of forms, such as decomposition diagramming and flowcharting, is aimed at understanding the activities and tasks that comprise a business function and how data flows between work units. Data modeling typically employs entity-relationship diagramming to represent and define the data used within an area being studied. Another useful technique, association diagramming, involves the use of matrices to depict interrelationships among various components such as organizations, entities, and processes. All these approaches are most effective in an innovation context if they are used sparingly, without going into heavy detail.

The use of data modeling is one of the principal differences between information systems-based approaches and those of other disciplines. Efforts that involve high-level, cross-organizational processes can employ data models to identify the information shared by various process activities and to highlight changes to the corporate data architecture. The data architecture, a representation of the information used throughout an enterprise, is independent of organization and function and ultimately guides the development and design of the enterprise's databases. Data modeling can help a firm determine when process innovation-effected changes to data influence the data architecture and identify the systems used by the other processes and business units. Data modeling is a requirement when the development of new systems is an absolute certainty following process innovation, although we have seen successful process innovation without data modeling, as in Federal Mogul's product development process.

Another distinguishing characteristic of information systems-based approaches is the relative novelty of measurement. No systems development methodology with which we are familiar measures the impact of a new system on process performance by comparing data gathered before and after implementation. As quality concerns continue to permeate Western corporations, we expect to see more attention paid to process measurement in systems development approaches.

Industrial Engineering

The task of industrial engineering is to design and redesign, through study, analysis, and evaluation, the components of man-machine systems.[11] Most often associated with process improvement in manufacturing processes, industrial engineering is in a state of transition. It is evolving from an emphasis on improving interaction between individuals and machinery through time and motion studies to the examination and optimization of systems needed to produce products and services. The new goals of industrial engineering sound much like those of process innovation. Whether the approach yields improvements or innovation may be an issue of scope rather than the concepts or techniques employed.

Industrial engineering has traditionally been concerned with the improvement of what was being designed or evaluated, with an emphasis on resource efficiency as a means of cost reduction. But the discipline offers many tools and techniques for analyzing and understanding existing, narrowly defined processes. It also considers human resource implications, at least at the level of the individual employee, a perspective that is absent from most other improvement approaches. In this regard, it resembles process innovation. Weaknesses of industrial engineering include the absence of information technology as a change enabler and the lack of focus on broad, cross-functional processes.

Among the many tools and techniques of industrial engineering is the process model, a work flow diagram that details the time and cost, overall cycle time, and idle time associated with given tasks or activities. Multiple activity charts are useful for differentiating between process activities performed by individuals and those performed by machines; network diagrams can be used to depict relationships among interdependent activities. Brainstorming and other group creativity techniques and mathematical models that can be used to optimize relationships between individuals and machines and even entire processes are other useful tools of industrial engineering.

SUMMARY

Inasmuch as understanding current processes is an essential element of all traditional process-oriented approaches, we look to these methods for tools and techniques that might be used to improve existing processes in an innovation initiative. Process diagrams, such as flow charts, are among the most useful of these tools. Traditional approaches use flowcharting to understand and communicate about processes. The cost buildup chart, another commonly used tool, can be used to reveal process bottlenecks and other problems, as well as process cycle time. In all cases, care must be taken to ensure that the tools and techniques selected fit the overall objectives of the innovation effort.

None of these traditional improvement approaches is likely to yield radical business process innovation. Although they may share characteristics with process innovation, all begin with the existing process and use techniques intended to yield incremental change; none addresses the envisioning, enablement, or implementation of radical change. They are most appropriately used to complement the components of the innovation approach described in this book.

All of the activities described thus far are prelude to the design and implementation of a new business process. In the next chapter, we examine the detailed design of new processes and consider the transition from current to future processes and the creation of process-oriented organizational structures.

Notes

[1]See, for example, H. James Harrington, *Business Process Improvement: The Breakthrough Strategy for Total Quality, Productivity, and Competitiveness* (New York: McGraw-Hill, 1991).

[2]Robin Cooper and Robert S. Kaplan, "Profit Priorities from Activity-Based Costing," *Harvard Business Review* (May–June 1991): 130–135.

[3]H. Thomas Johnson and Robert S. Kaplan, *Relevance Lost: The Rise and Fall of Management Accounting* (Boston: Harvard Business School Press, 1987).

[4]Robin Cooper and Robert S. Kaplan, *The Design of Cost Management Systems* (New York: Prentice-Hall, 1991).

[5]Ibid.

[6]Rainer Famulla and Ruth Schindler, "Business System Redesign: A Sensible Restructuring Tool," *American Banker* (August 2, 1990).

[7]Harrington, *Business Process Improvement*. For a less thorough but similar approach, see George D. Robson, *Continuous Process Improvement* (New York: Free Press, 1991).

[8]For example, Ernst & Young's Navigator System Series, [SM] a full life-cycle methodology for developing information systems, includes identification of process problems and outlines short-term improvement opportunities.

[9]James Martin, *Information Engineering: Books 1–3* (Englewood Cliffs, N.J.: Prentice-Hall, 1989).

[10]IBM internal document, "Using Information Strategically," November 1990.

[11]Gavriel Salvendy, ed., *Handbook of Industrial Engineering* (New York: John Wiley, 1982).

Chapter 8

Designing and Implementing the New Process and Organization

Ironically, there is less to say about the design phase of process innovation than about the activities that lead up to it. The design activity is largely a matter of having a group of intelligent, creative people review the information collected in earlier phases of the initiative and synthesize it into a new process. There are techniques for facilitating the review process, but the success or failure of the effort will turn on the particular people who are gathered together.

The choice of participants for a new process design team should be governed by both design and implementation considerations. A balance must be struck between team members who can deliver the most creative and innovative process solutions and those who can help to ensure that they are implemented.

Although in most organizations, the same individuals who participated in the process selection and visioning phases will participate in the design phase, it is particularly important that key process stakeholders feel their interests are represented during the latter phase. Stakeholders who should participate on the team during the design phase include heads of key functions intersected by the process, key general managers with operational responsibility for the process, suppliers of important change resources (e.g., the IT, human resource, and financial functions), and process customers and suppliers, both internal and external. It is worth noting that a number of consumer goods manufacturers, among them Procter & Gamble, Scott Paper, and Black & Decker, invited a few large retailers to participate in the redesign of their order management processes for quick response and continuous replenishment. In fact, exclusion of important customers such as Wal-Mart from the design phase would be virtually unthinkable. Inclusion of these stakeholders may make a design team unwieldy, but time lost achieving large group consensus can be recovered by shorter implementation times.

Our discussion here covers the implementation of new process designs, including requisite intermediate steps, and the implementation of new organizational structures and systems designed to reinforce and support new ways of working. Key activities in the design phase are listed in Figure 8-1.

BRAINSTORMING DESIGN ALTERNATIVES

Design innovation is best accomplished in a series of workshops, and brainstorming is an effective means of surfacing creative process designs.[1] By brainstorming, we refer to any group facilitation technique or practice that encourages participation from all group members, regardless of their roles and relationships within the organization. Emphasis during brainstorming sessions should be on creativity and idea generation, and a nonjudgmental atmosphere is essential. Any idea should be fair game, and participants must feel certain that they can share their thoughts without risk. The objective of brainstorming sessions is to develop creative, but pragmatic new process designs, taking as input the process vision, change enabler, and benchmark knowledge developed in the earlier phases of process innovation.

Graphic representation of a process design can be extremely helpful in understanding process flows. The use of computer-based tools for design display and simulation is discussed in a later chapter, but less technological approaches can also be very useful.

Figure 8-1   Key Activities in Designing and Prototyping a New Process

- Brainstorm design alternatives
- Assess feasibility, risk, and benefit of design alternatives and select the preferred process design
- Prototype the new process design
- Develop a migration strategy
- Implement new organizational structures and systems

Firms have successfully employed large whiteboards and large pieces of colored paper and string affixed to walls. Most computer-based tools use a rigorously defined set of symbols to represent different process entities, but these are not essential. The primary purpose of the graphic display is communication and recording, and any consistent set of easily understood symbols will suffice.

It is often useful to define the new process in an iterative fashion, with greater detail at each successive level (see Figure 8-2). To begin with, a high-level flow of the overall process should be created. This should not be difficult, given a well-articulated process vision. At the next level of detail, each subprocess can be described with roughly the same thoroughness as was used in describing the general process in the first iteration. Finally, each major activity should be described in terms of such factors as who will perform it, the information needed to carry it out, and so forth. For the sake of coherence and consistency, different graphic icons should be used to portray each level.

Figure 8-2   Levels of Process Design

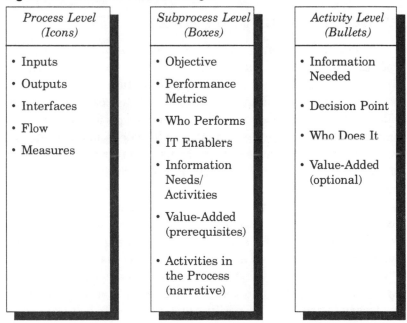

Process Level *(Icons)*	*Subprocess Level* *(Boxes)*	*Activity Level* *(Bullets)*
• Inputs	• Objective	• Information Needed
• Outputs	• Performance Metrics	
• Interfaces	• Who Performs	• Decision Point
• Flow	• IT Enablers	• Who Does It
• Measures	• Information Needs/ Activities	• Value-Added (optional)
	• Value-Added (prerequisites)	
	• Activities in the Process (narrative)	

ASSESSING FEASIBILITY, RISK, AND BENEFITS

The minds of design team participants should be full of information as the design activity begins. Participants should be well versed in the high-level vision for the process, including performance objectives and attributes, and be aware of opportunities and constraints presented by likely enablers, both technological and human/organizational, of the target process. They should be familiar at least with the broad performance parameters of the current process. Analysis of the current process need not be rigorous, but high-level aspects of the analysis should be inputs to the design deliberations.

Brainstorming sessions usually produce a number of design alternatives, which must be submitted to feasibility analyses to evaluate their relative benefits, costs, risks, and time frames. The new design and current state must be compared in terms of structure, technology, and organization to fully understand the implications of each alternative. The results of these analyses provide the basis for selecting the optimum design.

PROTOTYPING THE NEW PROCESS

Developing a prototype is a way to simulate and test the operation of a new process. It is an iterative process in which the fit between new process structure, information technology, and organization is refined and re-refined. A prototype might be considered the analog of a scientific experiment performed in a laboratory setting. Our definition is a small-scale, quasi-operational version of a new process that can be used to test various aspects of its design. This kind of prototype is known as an organizational prototype.[2]

There is no way to predict the organizational impact of a redesigned process and associated information technology with complete accuracy. The goal of the prototyping is to gradually shape the organizational environment or, alternatively, to revise the technology. Prototyping must be viewed as a learning activity by process designers and users alike. Many iterations may be necessary to achieve a proper fit; thus, the need to reiterate must not be viewed as failure.

A new process for taxpayer collections that was not prototyped by the Internal Revenue Service, although far superior to the old process and well supported by a new information system, had an unanticipated and negative impact on workers and supervisors. Workers were forced to spend all their time at their terminals or on the phone, supervisors to spend all their time monitoring workers. Both resigned in large numbers, forcing the IRS to reevaluate the new process and job roles. Eventually, it was necessary to spend more than a million dollars to modify the systems to support a team-based rather than individual process design.

Prototyping can itself be viewed as a series of phases that yield increasing degrees of tangibility (see Figure 8-3). Computer simulation, one of the techniques discussed later, is a kind of limited process prototyping, beyond which it may be reasonable to create a paper-based information test of the process. In subsequent phases, the prototype might be taken to a stand-alone process test, using personal computers for information support, and interfaces to other processes and existing information systems might then be added to it. The ultimate prototype would include all enabling technologies, skills, and organizational structures. Each phase helps refine the process design and the information required to support it; taken together, these phases help reduce implementation risk.

Figure 8-3   Levels of Process Prototyping

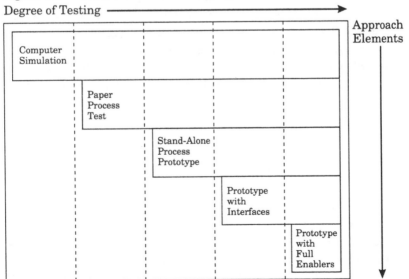

PHASES OF PROTOTYPING

A videotape that illustrates a new process can be useful for persuading customers to change interfacing processes and senior managers to fund implementation of the new process. Seeing a process on tape lends it a tangibility that paper reports and third-hand discussions seldom achieve.[3]

Clearly the organizational learning perspective needs to extend beyond prototyping. It sometimes takes a few cycles to arrive at an effective process design. IBM Credit Corporation, for example, has implemented a number of different process designs over the past five years. When desired objectives were not obtained, it tried again. In fact, one of the lessons managers at IBM Credit have learned is that process innovation never really ends. Having achieved tenfold improvements in cycle time and quote volume, the organization is attempting to push process performance into the field to achieve even greater benefits.

MIGRATING TO A NEW PROCESS

Having designed and tested a prototype process, an organization faces the considerable challenge of migrating from the current process environment to the radically new design. A full "cutover" may be difficult or impossible. If the new process involves customers, revenues, or valued employees, or if the process change will be highly visible internally or externally (and for what important process are these conditions not true?), the firm may not want to risk an abrupt transition. Alternatives to full cutover include phased introduction, creating a pilot, or creating an entirely new business unit.

A pilot is a smaller scale, but fully operational, implementation of a new process in a relatively small unit of the organization based on a particular geography, product, or set of customers. Although pilots are often viewed as a means of testing a new process (or other type of intervention), the goal should be to achieve success rather than merely objectively test. Thus, the unit selected should be the one most capable of achieving successful change.

One migration approach is to begin with a pilot and follow with a phased introduction. A firm might, for example, implement new systems capabilities and skills as they become available. A phased approach may be the most economically feasible, in that companies can derive some financial benefit from the process change earlier than might otherwise be possible; it is not, however, necessarily less disruptive than a full cutover. In fact, the sense of constant change and instability may be difficult for some employees to handle.

If constraints within the existing environment are too great, it may be desirable to create a new organization for the new process. This organization can run parallel to the existing one and be the locus of specific products, channels, or customers. The most prominent example of process innovation in the banking industry involved establishing a new organization. Midland Bank in the United Kingdom established an organization called First Direct to service retail customers without normal branches or other "bricks and mortar." The new bank employs innovative customer service processes that rely on the telephone and automated teller machines. It also makes extensive use of information technologies to identify patterns of customer behavior that reflect credit risk. Because its processes do not involve expensive real estate, First Direct can offer 24-hour service and higher interest rates that compete with Midland's more traditional banking organization as well as with other U.K. banks.

IMPLEMENTING NEW ORGANIZATIONAL STRUCTURES AND SYSTEMS

Firms and organizations today tend to be structured in a way that works against the success of their new process designs. Most organizational structures are based either on function or product, with little or no process orientation. Functionally organized firms have difficulty meeting customer needs seamlessly across different functions because no one "owns" the issue of how long it takes or how much it costs to fulfill customer requests. Certain key processes—typically new product development and order management—cross so many parts of the organization that the only manager to whom all their activities report is the CEO.

A firm organized around product structure has a difficult time ascertaining total business done with individual customers or "cross-selling" different products to the same customer. The latter problem is particularly pressing in banks, which, being organized around product lines (for example, many have a trust system for trust customers, a demand deposit accounting system for checking account customers, a consumer loans system for consumer loan customers, and so forth) are encountering severe problems establishing integrated customer databases.

A number of banks, including Continental and Chemical, are attempting to impose an "overlay" architecture that would enable a customer relationship manager to reach into the various databases in which customer information is contained to create a composite customer file.[4] But until the basic organizational structure is no longer product-driven, this activity will be contrary to the systems, reporting responsibilities, and culture of the bank.

Although the problem of rigid functional organizations is widely recognized, the proposed solution—to abandon any form of structure beyond the self-managing team—is frequently worse than the problem, or at least much less well defined.[5] We cannot imagine that real firms will abandon structure to the degree suggested by the set of concentric circles,[6] the orchestra,[7] or something else, or that they would be effective if they did so. We are, Elliot Jacques has noted, creatures of structure, and have been since the beginning of formal organization.[8] A more process-based, rather than post-structural, organization offers a powerful compromise between the need to maintain structure and the desire to adopt a flexible approach to the way work is done. Until a process-based organization is established, a new design for work cannot be considered fully implemented.

We do not recommend that processes become the only basis for organizational structure. Functional skills are important to a process orientation, as is concern for product management and the running of strategic business units. Most firms are well advised to adopt a multidimensional matrix structure, with process responsibility as a key dimension. An organization that wishes to benefit from a process perspective must be prepared to tolerate the well-known problems with matrix structures, including diffusion of responsibility, unclear reporting relationships, and excessive time spent in coordination activities and meetings.

There is another reason a purely process-oriented structure will not solve all structural problems. Firms such as IBM and Xerox, which are vigorously pursuing multiple process innovation initiatives, have made interesting observations about structure. Just as key activities can fall between the cracks of functions, important activities can fall between the cracks of processes, even broadly defined processes. If organizations converted fully to process-based structures, some future researcher would undoubtedly speak of the need for cross-process integration. Multiple dimensions of structure can help bridge the gaps created by a single structural dimension.

But only when firms adopt more process-based organizational structures will processes be managed in congruence with other aspects of the organization. Then, instead of cutting across the organization, process responsibilities will be a key focus of the organization. Process ownership, rather than constituting a "shadow" organization, as some companies have referred to it, will be a primary dimension of reporting and performance measurement.

In effect, a process-based organizational structure is a structure built around how work is done rather than around specific skills. A number of organizational theorists have argued that organizations need to reduce their levels of hierarchy and adopt action-based rather than formal structures.[9] A process-based structure combines an action orientation with some degree of formal structure.

There is a harbinger of the effectiveness of process-based structures in the way several Japanese automobile manufacturers approach product development. Kim Clark and Takahiro Fujimoto have described the concept of the "heavyweight product manager" in these firms.[10] This role has coordination responsibility for product development from concept to market, across such functional areas as engineering, manufacturing, marketing, and sales. Although relatively high in the organizational hierarchy, the heavyweight product manager exercises relatively direct influence over the activities of engineers at the worker level.

Although Clark and Fujimoto do not take an explicit process perspective, the heavyweight product manager role might be equated to ownership of the product development process. Only through such a role can companies develop products as complex as the modern automobile with speed, efficiency, and design and marketing coherence.

Yet we know of no Western organization that has made radical strides toward a process-based structure. Most companies that have undertaken substantial process innovation initiatives have simply imposed process management as an additional dimension of structure—on top of the existing dimensions—assigning process ownership to managers who may also have functional and/or product responsibilities. In almost none of these firms has process responsibility been accorded organizational legitimacy.

The hesitation to move to process-based structures goes beyond simple unfamiliarity with process as an organizational unit. When asked why they were reluctant to move toward process-based structures, process-oriented managers offered, among other reasons:

- concern that the level of organizational change from process innovation is already high and converting to a process-based organization might constitute too much change;

- fear that if functions are no longer the primary basis of organization, functional skills will be lost; and

- a belief that process is an unstable basis for organization because processes change more quickly than functions.

Only the first objection, we believe, has any real merit. Although it may be more difficult to preserve strong functional skills in a process-based organization, it seems quite possible. Marketing research skills, for example, are just as important to a product development process as they are to a marketing function. But because such skills may be spread throughout several processes, there must be structural or systemic approaches to nourishing and preserving them. With regard to the stability of processes, the premise of the objection may be valid, but not the conclusion. The fact that processes may change often to better meet customer needs is not an argument for abandoning them as a basis for organization. In fact, many organizational theorists have argued that a more dynamic basis of organization is exactly what is needed.

Only the concern that changes in organizational structure— on top of organizational changes imposed by process innovation— might be too much change all at once is well founded. Companies can begin their process innovation initiatives with shadow process organizations, and then slowly migrate toward a full process-based structure. As Beer et al. have observed, a congruent organizational structure is important, but many companies too quickly create new structures in major change programs rather than address organizational development and individual behavioral change first.[11] The timing and management of the change considerations involved in moving to a process structure are explored further in the next chapter.

BEYOND PROCESS DESIGN

The innovation approach outlined in the last several chapters is only the first step toward full-scale implementation of new processes. The complete innovation cycle is depicted in Figure 8-4. Innovation and the creation of a new organizational structure must be followed by detailed systems design, development of new performance measurement systems and skills, and systems construction and deployment. A fully implemented process innovation occurs over several years, and although our approach is outlined as a sequential process, it should, in fact, be executed in a highly iterative fashion. Rigid partitioning of the activities will not yield the maximum benefits of innovation.

Figure 8-4 The Process Innovation Cycle

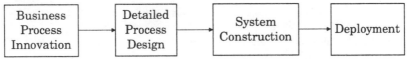

Companies involved in process innovation initiatives often act as if the most difficult aspect of the effort is over when the new process design has been developed. But realization of process innovation benefits is not, at this stage, a fait accompli. Like those of any other innovation, the benefits of process innovation appear gradually in response to active management. A number of companies, including Sea-Land[12] and Du Pont, employ specially designated teams to follow innovations well after implementation in order to determine whether benefits from IT-enabled business changes are being realized. Although such teams typically do not take a process perspective, the concept of benefit management and post-implementation assessment would be facilitated by process thinking, with its emphasis on measurement and outputs.

Notes

[1] Gini Graham Scott, "Making Effective Decisions Through Brainstorming," *Manage,* 41:3 (November–December 1989): 34–36.

[2]Dorothy Leonard-Barton, "The Case for Integrative Innovation: An Expert System at Digital," *Sloan Management Review* (Fall 1987): 7–19 and "Implementation as Mutual Adaptation of Technology and Organization," *Research Policy* (February 1988): 251–267.

[3]I am grateful to Professor James Cash of Harvard Business School for this suggestion.

[4]Stuart Madnick and R. Wang, "Evolution Towards Strategic Applications of Data Bases Through Composite Information Systems," *Journal of MIS*, 5:2 (Fall 1988): 5–62.

[5]See, for example, Steven Dichter, "The Organization of the 90's," *McKinsey Quarterly* 1 (1991): 145–155.

[6]Shoshana Zuboff, *In the Age of the Smart Machine: The Future of Work and Power* (New York: Basic Books, 1988).

[7]Peter F. Drucker, "The Coming of the New Organization," *Harvard Business Review* (January–February 1988): 45–53.

[8]Elliot Jacques, "In Praise of Hierarchy," *Harvard Business Review* (January–February 1990): 127–133.

[9]See, for example, Robert G. Eccles and Nitin Nohria with James D. Berkley, *Beyond the Hype: Rediscovering the Essence of Management* (Boston: Harvard Business School Press, 1992).

[10]Kim B. Clark and Takahiro Fujimoto, *Product Development Performance: Strategy, Organization, and Management in the World Auto Industry* (Boston: Harvard Business School Press, 1991).

[11]Michael Beer, Russell A. Eisenstat, and Bert Spector, *The Critical Path to Corporate Renewal* (Boston: Harvard Business School Press, 1990).

[12]Michael L. Sullivan-Trainor, "IS Keeps Sea-Land on the Move," *ComputerWorld* (April 29, 1991): 67–76.

Part II

The Implementation of Innovative Business Processes

Introduction

The previous section of this book explained the phases of a process innovation initiative. In this section, approaches to the successful implementation of process change that cut across these different phases are described. Though process innovation is an appealing way to bring about substantial operational improvements in a firm, it is difficult to accomplish. Any factors that can facilitate successful implementation should be explored. As elsewhere in the book, both human/organizational and technological factors are considered.

In Chapter 9, process innovation is viewed as a type of large-scale organizational change. Factors that contribute to the successful management of change in a process context are presented and discussed in considerable detail. Such issues as understanding process-oriented change, structuring the change, and managing the change from day to day are described. To better illustrate these issues in their full organizational context, a detailed case study at Digital Equipment Corporation is presented.

In Chapter 10, a very different set of implementation issues is addressed. These involve the role of information technology in facilitating process change. We have already considered the role of this technology in enabling innovative process designs. It is even more common, however, to use information systems to implement already-designed processes. This chapter takes a somewhat broader view of implementation, and addresses potential roles for IT throughout the entire set of phases for process innovation. While information technology is not as important a factor in implementing process innovation as organizational change, it can provide significant benefits and shortcuts to a fully operational process design.

Chapter 9

Process Innovation and the Management of Organizational Change

Business process innovation, despite its promise of radical competitive benefit, is still a rare phenomenon in the corporate world. Growing numbers of executives are aware of it, but very few have undertaken serious process innovation initiatives. The reason for this is simple—business process innovation requires abandoning comfortable old ways of doing business. It necessitates thinking about organizations and organizational boundaries in new ways that involve major, large-scale organizational change. In short, business process innovation involves radical change, and individuals and organizations cannot always be expected to embrace such change.

Organizational, not technical, barriers present the major challenges in process innovation efforts. "I often get managers coming to me complaining about my inability to integrate the technology," explained the chief information officer of a large aerospace and defense contractor. "And I always tell them that if it is really important to them, I can get any two boxes talking to each other by next week. They just walk away because they don't want to face the organizational issues."

Successful implementation of process innovation depends on consciously managing behavioral as well as structural change, with both a sensitivity to employee attitudes and perceptions and a tough-minded concern for results. This combination is difficult to come by. We are familiar with many firms that have readily identified new processes that promise radical performance improvement, but have been unable to implement them because they could not effect the required organizational change. Managers who perceive the need for change, but are blocked by one of a number of common pitfalls of change management, are likely to become frustrated, having seen the possibility of company-saving process innovation and been unable to realize it.

In this chapter, we explore critical factors that must be understood to manage effectively the organizational change associated with process innovation. First, we examine the nature and characteristics of process innovation as organizational change. Managers who would pursue change must understand the level and type of change required. Next, we outline a structure for process-innovation change management. Structural elements include change roles and qualifications for assuming these roles, and types of process change teams required. Finally, we discuss the actual management of change, including timing, scope, and pitfalls.

Because many issues of organizational change are specific to an organization's context and culture, the only way to fully explore process innovation as organizational change is through a case study. We selected Digital Equipment Corporation's Distributed Systems Manufacturing Group, an example that involves changes in several key processes related to manufacturing and logistics. We have supplemented elements presented in a Harvard Business School case study[1] with in-depth interviews of managers involved in the case.

THE DISTRIBUTED SYSTEMS MANUFACTURING GROUP

The Distributed Systems Manufacturing (DSM) Group, a rapidly growing producer of network products within the Computer Systems Manufacturing Group of Digital Equipment Corporation (DEC), employed 1,100 people distributed among its Boston-based headquarters, four engineering groups, and three plants located in Maine, Puerto Rico, and Ireland. Demand for its approximately five hundred products was increasing rapidly in the mid-1980s among value-added resellers, distributors, and direct and other external customers, as well as within DEC. With annual growth rates of 40% to 60% anticipated for the period 1987–1989, DSM was expected to generate almost $1.5 billion in revenue by 1990.

In 1985, the former materials manager for Computer Systems Manufacturing became head of DSM, charged with doubling return on assets and realizing a 10 point annual improvement in margins over five years, while keeping head count flat.

The new manager and his staff, some of whom were hand-picked, assembled a five-year strategic plan predicated on making dramatic improvements in three key processes: managing supply and demand, manufacturing, and new product development. The plan, which called for manufacturing cycle time and time to market to be reduced by 50% and new product introductions to be tripled, was based on a vision of a "virtually integrated enterprise" termed the "End Point Model" (see Figure 9-1).

A nine-month planning process commenced with assessments of competitors and future customer requirements that emphasized the importance of reducing manufacturing cycle time in order to meet customer demands for responsiveness. Reducing the time from vendor shipment of parts to delivery of products to customers hinged on tighter linkages among DSM groups and between DSM and its customers and vendors, of which there were thousands worldwide by 1985. Given this geographical dispersion, the only way to achieve the desired level of integration was to create a "virtually integrated enterprise" (enterprise in this case meaning not just DSM or even DEC, but the network or partnership of all firms involved in delivering completed products to customers) based on networked systems.

The DSM team developed an aggressive five-year plan (Figure 9-2) consisting of programs and activities at the group and plant levels that were, according to DSM's new manager, both

Figure 9-1 The DSM End Point Model for Enterprise Integration

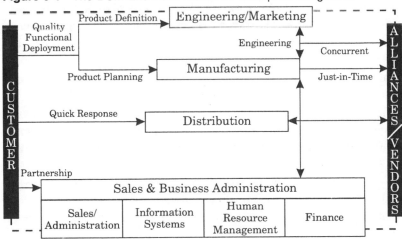

Figure 9-2   End Point Plan

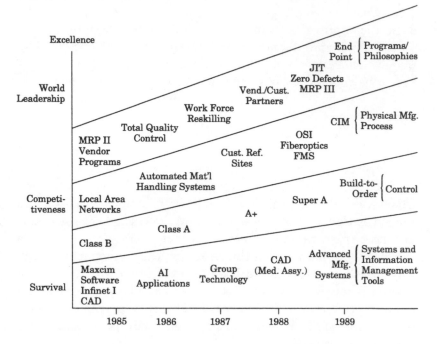

"IT and culturally intense." A systems and information manage-ment-tools component called for the implementation of computer-aided design (CAD), computer-integrated manufacturing (CIM), artificial intelligence (AI), group technologies, and other ad-vanced manufacturing systems, many of which had significant impacts on how people in the organization worked. Another sec-tion that dealt with integrated logistical processes, including plant-level materials handling and information flows, and with product delivery-time reduction called for DSM to reduce its cycle time from 40 weeks to 15 days over the five-year period. Measures associated with manufacturing resource planning (MRP II) cer-tification, a major objective of the plan and a means of achieving organizational discipline and integration, were intended to meas-ure DSM's progress toward achieving build-to-order capabilities in 1990.

When DSM's new manager left the organization in 1988, he left a legacy of dramatic achievements in both process perfor-mance and consequent financial results (see Figure 9-3) with which most any firm would be pleased.

Figure 9-3   DSM Performance Improvement, 1985–1988

Process		Financial	
Cycle Time	–52%	Inventory	–60%
Time to Market	–30%	Revenue	+2X
Quality	+95–99%	ROA	+25 points
Delivery	+85–95%	Margin	+25 points
Product Intros	2.5X	Headcount	Flat

UNDERSTANDING PROCESS INNOVATION AS ORGANIZATIONAL CHANGE

As a form of organizational change, process innovation is characterized by simultaneous dramatic change in many structures and systems. To manage successfully across all these fronts, managers must be aware of the nature of process innovation as organizational change, the likely sources of problems and resistance, and the areas in which their attention is likely to be required. Specifically, process innovation-oriented change must be understood in terms of five characteristics: the overall magnitude of change required, the level of uncertainty about change outcomes, the breadth of the change across and between organizations, the required depth of penetration of individual attitudes and behaviors, and the duration of the change process.

Process innovation and the organizational change it occasions are equivalent to radical surgery. They are the triple bypass operation, the removal of a large malignancy, the transplanted organ. Given the magnitude of the risk and reward, process innovation should be undertaken when more moderate business improvement approaches have failed or will clearly fall short of the need. Process innovation can only be accomplished when the leaders of an organization believe and can demonstrate that current modes of operation are a threat to the survival of the company.

Pain, hope, and uncertainty.   The pain of the current state must be visible not only to outsiders, but widely accepted by senior managers within an organization. Often, consultants, the business press, and a handful of executives are aware that problems with key processes threaten a firm's long-term survival, but are unable to convince the majority of the firm's managers that radical corrective action is required. In one automotive firm, for example, "everyone" knows that unless the product development cycle is cut in half, Japanese competitors will continue to steal market share, yet key executives point proudly to the volume of new models and deny that development cycle time is a critical issue. The firm's chairman has mandated cycle time reductions, but no one has taken responsibility for making the change happen.

As we will point out later, the sense of organizational "pain" can be manipulated to some degree by senior management, but unless such pain is widely felt and acknowledged by key members of the executive team, any process innovation initiative is likely to be sandbagged or sabotaged. The trick, of course, is to convince an organization of the need for change before a crisis occurs—while there is still time to do something about it. The pain associated with maintaining the status quo must be seen to exceed the inevitable and very considerable pain associated with the transformation to a new way of operating.

On the other side of the motivational coin are hope and positive expectations. Some process innovation initiatives are motivated by hope alone, but most involve a combination of pain and hope. Hope for greatly improved operational and economic performance must be held out to the organization by its leader in the form of a vision. To hold the attention of management, and thus the rest of the organization over the long term, this vision must describe changes in ways of operating as inarguably strategic imperatives. Statements about the objectives and attributes of the future process state are inherent in our definition of a vision.

The DSM manager was aware that he had been handed a difficult assignment when he was asked to make dramatic improvements in the organization. First, it would be difficult to effect change in a successful company (which Digital was at the time), particularly in an organization experiencing a period of rapid growth, which lent no immediate impetus for changing processes radically. But with success, the organization had become complacent, and the manager knew that to achieve the business objectives he had been handed he would have to wake up a sleepy manufacturing organization.

There were chinks in DSM's armor that the manager could exploit. For one thing, the organization had placed next to last in internal and external benchmarking studies of manufacturing performance. The manager used this ranking, together with other data gathered in the assessment process, to create impetus for change. The message he sent to his organization was that it had a major problem, and that cooperation from everyone was needed to solve it. DSM's lackluster report card provided the pain; the hope the manager and his team held out was the End Point Model vision of manufacturing excellence.

Uncertainty about the end state is as characteristic of process innovation change as it is of other types of major change in corporations.[2] It is neither possible nor desirable for a leader or anyone else to define at the outset precisely how a transformed company will operate. Process objectives and attributes must be stated in advance, but the process that emerges and the benefit that accrues to it cannot be anticipated—the scope and degree of change are too great. Second- and third-order effects of a given change or set of changes will inevitably lead to further unforeseeable insights and adaptations.

Major change of any kind, even change perceived to be positive, produces anxiety and resistance. Uncertainty about the end state heightens this reaction. To quell anxiety and overcome resistance, a successful change leader must not only foster awareness of the need to change, but also create realistic, positive expectations about the outcome. The manager must draw a bold outline of a compelling future state and be able to fill in enough detail to convince others that to follow is to gain higher ground.

Coordination and cooperation. Process innovation, by definition, spans traditional organizational boundaries. Propelled by mounting competitive pressures and a need to reach new levels of organizational performance, leading companies have realized that functional management and incremental functional improvement are no longer sufficient; in fact, process changes occurring today span inter- as well as intraorganizational boundaries.

We have noted that the larger and more diverse an entity, the more difficult it is to effect and manage change. That process innovation requires increased coordination and cooperation across both internal and external interfaces places unprecedented demands on the change management process, and requires, in addition to time and money, skill, determination, persistence, and patience on the part of the leaders and managers involved in the change.

Major changes that have occurred in many firms' manufacturing processes over the past decade illustrate the requirements for new levels of internal and external coordination and cooperation. Just-in-time (JIT) inventory management is a case in point. To speed the flow of goods through their factories, manufacturers have reduced the work-in-process (WIP) inventories that buffered different parts of the manufacturing process, obscuring process imbalances and quality problems and clogging the flow of goods. Maintaining a steady flow of goods in a bufferless state requires tighter coupling not only across the steps in the manufacturing process, but also with the production and materials planning organizations. To provide scheduling stability, the linkage between forecasting and production planning must be tightened and their respective cycle times reduced. This tearing down of internal functional walls is known as "internal JIT," or "continuous flow manufacturing."

In order to fully implement JIT, manufacturers must also control the quality and flow of factory inputs, which means major changes in relationships with vendors and tighter integration of production processes with vendors' order/delivery cycles. Behavioral and structural changes of equal magnitude are likely to occur in other processes, from product development to billing/payment.

Cultural and paradigm shifts.   Creating tighter internal linkages between functions entails not only changing structure, but also bridging cultural differences and upsetting traditional power balances. In order management processes, for example, the field sales personnel also become credit checkers, manufacturing expediters, and logistics analysts. Functional managers must trust that employees have the capacity to gain the experience needed to perform their tasks. Functional organizations, even in companies with strong corporate cultures, tend to have their own operating characteristics and culture based in part on the nature of their core discipline and the kind of people that discipline attracts, in part on the style of the leaders who have shaped the organization, and in part on the organization's place in the functional pecking order of the company. Sometimes these barriers are so great that it is easier to integrate with customer or supplier processes than with other functions within the same company.

Participants in this change pursue it not out of newfound altruism or kindheartedness, but because costs of traditional ways of doing business are so high and the benefits to be reaped from change so dramatic. Interfunctional and interorganizational trust are desirable generally, but they are only essential in the context of process innovation.

Another dimension of the organizational impacts inherent in DSM's End Point Model is the necessary shift in long-standing paradigms about work and organization. The notion of a multifirm "enterprise" conveyed an expanded concept of organization, one that included DEC's customers and vendors. Individuals and groups within DSM worked through the implications and new behaviors required as a result of other equally profound changes in assumptions. The organization's fundamental values had to shift from emphasizing what was good for the firm to emphasizing what was good for the enterprise. A strong functional and job focus had to be replaced by a cross-functional, cross-organizational process orientation. Informal systems and lax controls had to be replaced by a disciplined, heavily measured set of formal process tasks.

Change incurred by process innovation is not only broad, but deep, extending from the visions of managers to the attitudes and behaviors of the lowest-level workers. As with other types of major organizational change, the behavioral aspects of process innovation equal the structural aspects in magnitude and importance, and often are more difficult to address. Its significant behavioral component makes process innovation-based change qualitatively different from other forms of large-scale restructuring; it runs deeper than even major acquisitions or asset liquidation. Organizational charts come and go, but changes in the way a business and its employees operate on a daily basis cannot be prescribed on paper.

Behavioral change.   Successful leaders of transformational restructuring understand that changes in mental models, attitudes, values, and, ultimately, behavior are the foundation for successful implementation of these changes in operational and management structures and systems. They also understand that changes in mental models must precede changes in behavior. Greater cooperation between functional groups is an example of the kind of behavior that is almost always required to improve the performance of major organizational processes.

Too often, companies underrate the importance of developing new behavioral competencies and fail to devote sufficient time, effort, and resources to the task. Successful process innovation leaders see new behavior as an integral part of the vision of the way their organization will operate in the future. Recognizing that change in the behavior of individuals and groups is the foundation for successful institutionalization of structural process changes, they devote explicit attention to developing new behavioral competencies. Training to promote, and constant reinforcement of and informal rewards for, new behaviors are essential during the period of initial organizational learning.

Necessary behavioral changes are made even more difficult by the fact that process innovation cannot be a fully participative process. Opinions can be solicited and objectives and progress communicated, but regardless of the philosophical value of participative cultures, not every worker can materially contribute to a process design that affects thousands of workers. Managers of innovation initiatives must understand the burden that this inevitably top-down approach places on them.

Many of the programs associated with the DSM change effort—for example, work force reskilling, customer reference site, and Class A MRP II certification—had a significant impact on skills and jobs. Because creating a motivated work force, empowered through access to information and IT-based tools, was a conscious goal of the End Point plan, extensive investments were made in employee training and education to support the deployment of advanced systems and technology within the organization.

Duration of change.   A final aspect of process innovation as organizational change that must be understood is the duration of the change. Change of the magnitude we are describing usually requires a minimum of two years to plan and implement in a given business unit, and we have seen it take as much as five years. The multidimensional scope of transformational change is an obvious driver of the time requirement. It is simply not possible to plan and implement major changes in multiple organizational

systems and structures overnight. Moreover, successful implementation of transformational structural change is dependent on
changes in beliefs and behaviors, which take time to cultivate.

Managers must be aware of the receptiveness of their organizations to either new or ongoing change initiatives, and adjust
their plans accordingly. When process innovation is infeasible
because an organization is "burned out," the wise manager resorts
to a period of stability and process improvement. Many employees
at a major telecommunications firm, for example, viewed process
innovation (or reengineering, as it was called there) as yet another
buzzword meaning loss of jobs. It was perhaps the fifth major
change initiative in seven years that had led to major staffing
reductions. As a result, observed one manager, "whenever our
people hear the word 'reengineering,' they run for cover." Such
an environment is clearly not conducive to change; unless a firm
is facing short-term disaster (which this firm was not), a less
radical improvement program would better suit the organizational mood.

THE STRUCTURE OF PROCESS-ORIENTED ORGANIZATIONAL CHANGE

Internal organizational aspects of a process innovation effort
are central to the success of the overall project. Appropriate project organization is important not only to the execution of an initiative, but also to the change management activities described
above.

Given the risks and rewards inherent in process innovation,
a firm must establish a change structure within which specific
initiatives can be successful. Such a structure specifies key roles
and responsibilities for change tasks, the types of teams necessary
to design and implement new processes, and the composition and
skills of the individuals and teams who will participate. The establishment of this structure is, of course, most difficult for the
first process initiative; subsequent initiatives can benefit both
from the established structure and from learning about how it
performed the first time.

Roles and Responsibilities

Executive participation is important for communicating management support for and belief in an innovation project to the

entire organization. Organizational change associated with process innovation can only be driven top-down. It requires the commitment and involvement of the entire executive team, but ultimately it must be driven by a dynamic executive who serves as the focal point throughout the change process.[3]

Three qualities characterize successful transformational leaders: commitment and the ability to inspire, conceptual skills, and impatience for results, the last being coupled with an ability to deal with the "softer" behavioral aspects of change.

Commitment is an essential ingredient in all leaders. Daryl Conner's work has led him to believe that the executive sponsor's degree of commitment—commitment based on utter dissatisfaction with the status quo and a passionate belief in the new ways of operating articulated in the strategic vision—is the single greatest determinant of change management success.[4]

In a sense, DSM's group manager had been preparing himself for the position for a long time. He was given the opportunity to implement a new vision of a manufacturing company—a vision for which he had been working for three years and which he had documented for his boss in a white paper called "Materials Architecture." The title was somewhat misleading in that the paper dealt with an emerging vision of an integrated enterprise. The white paper was very well received and became a topic of discussion for a number of high-level corporate committees, which developed models or architectures describing how Digital might do business differently. By 1984–1985, the paper had evolved into a document titled "Corporate Manufacturing/Materials Architecture."

When the group manager, fresh from these experiences, accepted the leadership of DSM, he decided to emphasize improved performance over architecture. The document did, however, serve as an input to the planning process for the DSM executive team. The passionate commitment and almost religious zeal of the group manager and his strategic planning manager galvanized the group and led it to profess total ownership of the End Point Model it had developed in 1985.

Associated with this deeply felt commitment to a new operational vision is the capacity to inspire and incite. The words "charismatic" and "magic," used by David Nadler, are often associated with individuals who have this ability. During the political and social turmoil of the 1960s, a number of leaders in this mold, including John Kennedy and Martin Luther King, took center stage. In recent years, Mikhail Gorbachev and Boris Yeltsin have been the most obvious and notable transformational leaders on the world stage.

Although researchers have tended to focus on the emotional qualities of transformational leaders, major change efforts require more than commitment and emotional impact. The transformational leader is the initial owner, if not creator and primary proponent, of a change vision, and hence must be able to conceptualize and articulate totally new ways of structuring and operating a business.[5]

Successful business-unit change leaders are hard-headed, results-oriented executives who set high standards and remain dogmatic and insistent about business results. At the same time, they understand the importance of organizational change and have an ability to deal with its softer dimensions. In our experience, impatience for results *and* a realization of the conscious effort and time that is required to nurture new attitudes and behaviors in groups and individuals are hallmarks of successful transformational leaders. Process innovation leaders, especially at the business unit level, often must modify their own attitudes and behaviors to be consistent with the evolving norms.

But process innovation requires more than a single, charismatic leader. Because the requisite leadership functions range from originating a process innovation initiative, to making it acceptable and necessary, to managing it through to completion, most major change situations involve some type of division of leadership labor. Daryl Conner has identified four major change roles, which are as follows:

- the change advocate, who proposes change but lacks sponsorship;
- the change sponsor, who legitimizes the change;
- the change target, namely, the individuals or groups that must undergo change; and
- the change agent, namely, the individuals or groups that must implement the change.

Ensuring that the right people fill each role is crucial. One of the most common mistakes made by companies embarking on process innovation is choosing the wrong sponsor.

At the level of discrete projects, executive participation helps foster an appropriate perspective on company strategy, essential for forging the link between project objectives and business strategy discussed earlier. The sponsor of process innovation activities plays a pivotal role in defining and driving change within a business unit. The individual of choice must possess the commitment and abilities of the transformational leader. But the political and organizational positioning of the process innovation sponsor is also of critical importance. Executive sponsors should be among the most senior managers in a company.

At Federal Mogul, for example, the president of one of three businesses, who was also serving as interim general manager of the business unit, sponsored an innovation project in the sealing products division. At a bank holding company, a vice chairman sponsored the first major change initiative, and at Xerox, process innovation sponsors are presidents of divisions. But in too many companies, a senior information systems or human resource executive assumes the sponsor's role.

The sponsorship of change must usually be broader than a single executive. Only when it perceives a unified front on the part of the executive team will the rest of an organization take a process innovation initiative seriously. It is also important that an innovation initiative in a business unit have at least the tacit support of corporate executives. A Xerox order-management initiative in the U.S. Marketing Group, for example, has the full support of the division president and his senior staff, as well as of the corporation's CEO and most of his direct reports. Not only must members of the executive team agree, they must also actively communicate to the rest of the business unit and to their own areas in particular the need for change and their commitment to the vision of change.

It is important that the organizations and people subject to change—the change targets—report to the executive change sponsor. It is the sponsor whose personal, political, and organizational credibility and positioning must exert the requisite public and private pressure on key individuals and groups. This might include replacing key managers, which is one of the single most powerful change tools available to sponsors.

Analysis of a few examples of strategic business processes, such as product development or order management, with a focus on the change targets of those processes, will suggest an appropriate sponsor of process change. Major changes in the product development process need to involve marketing, engineering, and production, at a minimum. Sales and marketing, finance, and distribution need to be involved in innovations in the order management process. The locus where responsibility for all the organizations involved in strategic business processes comes together is the head of the business unit or, occasionally, the chief operating officer. Thus, in most cases, the general manager is the logical and necessary sponsor of strategic process change at the business unit level.

What does this mean for the marketing executive who has a vision of a revitalized and restructured product development process? Many executives today have the resources and skills required to implement change. But that, by definition, makes them agents, not sponsors, of change. Business-unit marketing executives control some of the activities that need to change if the right products are to be brought to market more quickly. But that, by definition, makes them change targets. Such executives, lacking the power and authority to legitimize and sustain broad-based change, must seek out and develop appropriate sponsorship.

Functional executives may also be advocates of change. In order to perform this role successfully, they must help introduce the concept of process innovation to the organization and persuade other senior managers of its merits. An IS executive who believes strongly that IT-enabled process innovation holds great promise for significantly improving business results is playing the role of change advocate. In fact, the advocate role is most likely to be played by IS in the organizations we have studied. The most successful change advocates try to broaden the advocacy role, and to identify sponsors and targets, as quickly as possible.

Ultimately, it falls to the change leader or sponsor to create and maintain strong commitment and consensus among the executive team members with respect to the need and vision for change and the plans for creating radical improvement in strategic processes. Failure to achieve executive team consensus can prevent a business unit's process-innovation efforts from ever getting off the ground, and failure to maintain commitment and consensus will diminish the degree of change that is achievable and delay progress and the realization of benefits.

The risks and rewards, costs and benefits of process innovation will never be apportioned evenly. Redistribution of power and authority among functions and levels and even among suppliers and customers is an inevitable outcome of process innovation. To navigate these tricky waters during the planning phase requires skillful negotiation and judicious application of pressure and support. To build commitment, executive sponsors first need to determine the receptivity of key stakeholders to major change. A variety of tactics, from informal discussions and education to reconfiguration of the executive team, are available to them for doing so. During the implementation phase, sponsors will have more formal accountability mechanisms at their disposal for ensuring that commitment is maintained. As part of the implementation plan, executive accountability for results must be assigned and progress carefully monitored through the tracking of milestones and targets for key measures. Ultimately, new reward structures will have to be established for executives.

A strategic-planning process is useful for building consensus about the external threat and developing and strengthening commitment to the vision for change. Offsite meetings might be conducted by external facilitators to solidify consensus and refine the vision. Once buy-in and consensus are established among executive team members through participation in refinement of the vision, including identification of strategic processes and the setting of objectives, process owners can be assigned to key processes, and process innovation teams created to flesh out the vision and develop plans for meeting the designated objectives.

Organizations that undertake a number of improvement initiatives concurrently are well advised to establish an executive committee to oversee and coordinate projects and counsel and guide the individual project teams. It is incumbent on top management to affirm and reaffirm that such teams are empowered to make radical changes. We know of senior executives on one project team who remained unsure about one of the process innovations they were about to propose until they were able to meet with the executive sponsor, who reassured them that they were, indeed, empowered to propose such a change.

One of the most important roles in individual projects is that of process "owner"—in terms of the change roles described above, the senior-most manager among the change targets. The owner has ultimate responsibility for a process. The cross-functional nature of many key business processes dictates that the process owner be at a high enough level to ensure authority over the process and all its interfaces. There may be no logical candidate for this role; he or she may simply have to be appointed by the sponsor of the change initiative. Process owners need many of the same leadership skills that sponsors possess, but at a somewhat more operational level. They need to be able to pick up where sponsors leave off and turn visions into implementation plans and plans into results. But leaders at both levels must be able to create cohesive, well-functioning teams.

For processes that cut across multiple companies or organizations, joint ownership by managers from each of the units involved is essential.

Types of Teams

The objective of process innovation is to optimize the performance of the business unit as a whole. This is often accomplished by creating new team structures that span traditional boundaries. The teams formed may be at two levels, executive teams and process innovation teams. Executive teams within a business unit make key decisions about process innovation initiatives and review results. Their work might include selecting processes for redesign, forming the process innovation team, supplying strategic direction to a vision effort, and selecting from alternative designs. They may be composed of change sponsors, agents, targets, and advocates—all at the senior-most level.

Process innovation teams do the detailed work of process innovation, including gathering information for process selection, searching for benchmarks, identifying enablers, creating more detailed visions, defining process flows, and creating prototypes and transition plans. As noted, some firms create a single team with all these functions. Other firms have achieved good results by creating separate teams, with minimal overlap, one to analyze the current process and another to design the future process. Given the importance of change management concerns, some firms have established a third, usually somewhat smaller, team specifically devoted to organizational change-management issues. Unless such a group is created, a process innovation initiative risks starting change tasks too late, or neglecting some change topics completely. Again, overlap of change-management team members should be minimized to prevent overload. Although such a multiplicity of teams is expensive, the cost is quite low when compared to the potential rewards.

Given the rationale for such teams—to "increase coordination across functions and activities so that the performance of the whole is greater than the sum of its parts"[6]—it is not surprising that they almost invariably remain important ongoing structures after process design efforts have concluded. In addition to being a mechanism for developing implementable recommendations, well-functioning process-innovation teams model the way the enterprise as a whole will operate in the future. At the same time, they are an important mechanism for developing the new behavioral competencies needed to make the process innovation succeed.

Team members derive motivation to develop new individual and group skills and behaviors first by accepting that radical improvement in the performance of strategic business processes is required, and later by perceiving that effective cross-functional coordination and cooperation is at the heart of well-functioning processes. Explicit training and coaching in teamwork, group dynamics, and evaluating differences is often provided to both process-innovation and business-unit executive teams. New group behavioral values and guiding principles are in many cases developed by the business-unit executive team. Occasionally, this task falls to individual process-innovation teams. What is important is that the teams explicitly and consciously debate, discuss, and internalize these new values and integrate them into their group and individual behavior.

Ancona and Nadler have identified three important team-management processes—work management, relationship management, and external boundary management—that need to be explicitly mastered by executive and process innovation teams.[7] Work management and relationship management processes are essential because of the high degree of interdependence among process-innovation team members. Work management deals with how the group makes decisions, shares information, and coordinates activities, relationship management with issues related to degree of openness and cohesiveness, conflict resolution, and levels of trust. Development of group behavioral values falls into the latter category. The volatility of the external business environment and the likelihood that relationships with customers, suppliers, and/or channel partners will be directly involved in or affected by process innovation argues for explicit attention to the process of managing team boundaries.

Composition and Skills

Several criteria should be applied when forming a process innovation team. First, there must be a stakeholder analysis. A stakeholder is anyone with a vested interest in the process—a customer, a department manager, an employee involved in one of the process activities. Stakeholder analysis identifies individuals who are likely to be affected by a process innovation. Whenever possible, key stakeholders should be assigned to project teams.

Managers and other employees who are likely to "lose out" as process innovation redefines and reconfigures activities and responsibilities must be involved in the project to foster commitment to its goals. To the extent that it is logistically impossible for all stakeholders to participate on the project team, other avenues must be found for allowing individuals who are left out to contribute. Data gathering comes to mind. Whatever the avenue, the activity must be at a level that ensures support.

A firm seeking customer participation in a process innovation initiative should solicit long-term, high-volume customers with which it has had, or wishes to have, a close relationship. A large department store that was redesigning its accounts payable process, for example, invited several of its key vendors to participate in the design of the new process.

Cross-functional representation on project teams is essential to ensure that a variety of perspectives is brought to bear on innovation efforts. One manufacturer attempting to reduce cycle time for new product development, for example, assembled a team that comprised individuals from sales, product engineering, manufacturing engineering, and the affected plants.

Finally, a process innovation consultant can play a key role in innovation efforts. Such consultants, who may be either internal or external, will possess experience in improving and redesigning processes. In many companies that have formed business units dedicated to assisting with business process innovation— CIGNA, Continental Bank, and IBM for example—the responsibilities of the information systems function have been expanded to include business process redesign. Senior IS professionals, who usually have an organizationwide perspective and may be politically neutral with respect to given processes, are often well qualified to serve as process innovation consultants. Other enterprises have established entirely new groups to provide process innovation expertise and facilitate and guide the change process. Such groups also typically provide training in the data-gathering and problem-solving techniques that will be employed in the change effort, and are candidates for managing process interfaces and facilitating interprocess project communications. Firms hiring external consultants often ask them to train internal personnel so that process innovation work can eventually be brought in-house.

Just as sponsors must exercise care in assembling their executive teams, so process owners and executive sponsors must carefully configure each process innovation team. Organizational representation is a primary factor. Clear articulation of the scope of the target process will help define the appropriate make-up of process innovation teams. All change targets must be represented directly, both for their functional knowledge and to ensure buy-in.

Team candidates' characteristics and skills must be carefully weighed against the functional knowledge they bring to the table. Ideally, creativity and openness will be blended with sound business judgment and the ability to synthesize information from multiple sources. Team members must grasp strategic realities and possess an operational understanding of the business. In today's organizations this combination will most often be found in the middle to upper-middle ranks of a business unit.

Team members must be trained in the techniques that will be used in the course of the innovation effort. These include problem solving, process documentation, and group dynamics. Training can also reduce the time it takes for team members to begin to feel comfortable working together.

Important change resources should be represented on a process innovation team. This could include members familiar with information technology, organizational development and human resources, or even financial restructuring. Most of the firms we studied had representatives from the IT and HR functions, or employees who had previously worked in these areas, participate as members of the team.

Team members skilled in human resource enablement may also play a team facilitation role. Facilitators, who, like process innovation consultants, may be either internal or external, possess experience in group development. The facilitator's role is to provide guidance to and promote cohesion among team members.

In configuring his executive team, DSM's group manager ensured that all functions and sites were represented. He also included five major vendors that, being DEC customers, were considered stakeholders.

One of the hallmarks of the DSM team was its diversity. The group manager chose team members with an eye to their openness to change. In his search for such individuals, he identified several women, appointing DEC's only female plant manager at the time. Members from DSM's two other plants, in Puerto Rico and Ireland, lent further diversity to the team.

The cultural and gender diversity of the team was ultimately one of its great strengths and helped it create a plan for integrating a very diverse organization. In the same way that effective cross-functional participation can produce solutions no single discipline could develop, the unique perspectives that each of the individuals and groups brought to the table enriched both the process and the end result.

Team members learned to listen to one another. They challenged unspoken assumptions and broadened each other's perspectives. Time was devoted explicitly to the development of new values and to working out group processes and group dynamics. Meetings were held at all DSM sites to enable the team to experience the culture of each of the different locations. The team worked together and played together. At the end of nine months, it emerged not only with a plan, but with a strong group identity and cohesiveness, reshaped perspectives and values, new skills and behaviors, a commitment to its vision, and a belief in its ability to work together.

MANAGING ORGANIZATIONAL CHANGE IN PROCESS INNOVATION

Even with a high level of awareness of change issues in process innovation and a structure of teams and team members, change must still be managed carefully. Among the factors to be emphasized in the change management process are the timing of change, the creation of a sense of meaning and mission, the management of scope, communications about the change process, and the avoidance of common pitfalls.

The timing of process innovation, as we have noted, involves a delicate balance between acting early, when the need for change may not be widely appreciated, and acting late, when the organization is desperate. Waiting too long will put management in a reactive mode and necessitate drastic measures. Such forcible imposition of change might rupture the trust between management and employees and thus make it impossible to achieve common objectives of corporate renewal such as a highly motivated work force. The very jobs of senior management are often placed in jeopardy by such crisis situations. It is better to time the initiation of large-scale change efforts to accommodate a more proactive and positive approach,[8] but given that transformational change is usually driven by a clear if not immediate threat, organizations of necessity tend toward a revolutionary rather than evolutionary approach.

Duration.   A number of other factors influence the duration of transformational change in a business unit. One, varying degrees of readiness among the leadership in various groups within a business unit will influence the pace, and thus duration, of the change effort; a group's leadership must buy into the need for change before any progress can be made. Two, building commitment to and encouraging involvement in a change effort takes time; the make-up of a group at different levels and the need for individuals to adapt or be replaced must be managed with a balance between urgency and sensitivity. Three, change programs need to coexist with ongoing business operations; the pace and duration of change will be driven by the amount of change individuals and groups can absorb and still continue to function productively. No manager should expect that process innovation for a broad process can be fully designed and implemented in less than 18 months to two years.

Scope.   Scope management is important because the larger and more diverse the entity to be changed, and the greater the change sought, the more difficult the process of effecting and managing it. The difficulty of managing strategic change in large, diverse entities has implications for the way change efforts are bounded. In corporate settings, strategic business units tend to be viewed as logical boundaries for change efforts. Getting functions within a business unit to cooperate is difficult; broadening the scope of change makes it even more so. When these units are themselves large and diverse, it may be desirable to carve off smaller, discrete organizational units.

A company composed of product-focused divisions may be called upon to respond to a major customer's demand for the ability to place consolidated orders and receive consolidated invoices. How does a company get multiple autonomous business units, with different ways of doing business and different cultures, to agree on common business practices? Process innovation is one way, but it entails considerable organizational risk. Securing consensus on needed process change from strategic business units and corporate center management is difficult, but trying to impose a process developed in one division or multiple divisions can be very unpopular. Companies trying to integrate their operations both domestically and globally are being forced to address these difficult structural and change management challenges.

DEC, it should be remembered, did not try to restructure the whole company at once; rather it started with one group, DSM. This approach was consistent with the company's entrepreneurial and decentralized culture, and it had beneficial effects. Indeed, such a focused approach is critical to successful implementation of transformational change and is required at several levels. The first level at which focus is required is in the bounding of the change effort.

Most successful process-innovation efforts we have observed have occurred within the bounds of a discrete business unit within a larger corporate entity. Some companies begin their change initiatives in a business unit on the periphery of the organization, in a single plant or small division. One pharmaceutical firm we studied, for example, rather than begin process innovation in its core business, tested its concepts in a relatively small medical-instruments business. The first processes CIGNA selected were in the reinsurance division, one of its smaller units.

Given the risks associated with transformational initiatives, starting on the periphery makes sense, assuming that a company's competitive situation accommodates this luxury and that managers of other businesses will not discount successful results. Starting on the periphery enables a company to mitigate its risk by testing and refining concepts in order to gain experience with managing process change, including the development of executives and managers. This is consistent with our earlier conclusion about the difficulty of managing radical change across large, diverse organizations, and with Beer et al.'s findings that corporate revitalization typically starts in discrete organizational units such as divisions or business units, plants, or branches and "depends on successful renewal in many organizational units."[9]

Even within a discrete business unit, transformational change is best managed in segments, rather than all at once. Our obvious preference is for process segments. Whatever the focus of the renewal effort, not all parts of an organization will have equal impact on its successful implementation. Thus, it is rarely necessary to undertake major change in all parts of a business unit simultaneously. Eventually, major changes in one part will ripple through the whole system, and other parts will change as the organization seeks a new dynamic equilibrium. But it is important to begin with a focus on key areas of the business.

Even if a business unit wanted to tackle all major processes simultaneously, few have the financial or human resources to mount such a broad-based assault. At the outset of revitalization activities, executives and managers with the ability to lead major change are a scarce resource. The time of such skilled leaders as do exist is an even scarcer resource, and thus a gating factor in the scope of change.

Values, attitudes, and behavior.   Changes in values, attitudes, and behavior cannot be effected by a decision taken in the boardroom or mandated in a memo. New process designs must come from the top down, but implementation of these changes is best likened to cultivating a garden. The ground must be prepared row by row. Gains on the behavioral front must be made group by group, person by person, including the leaders themselves. Successful change efforts attend explicitly to the development of new attitudes and behaviors. Extensive training is often provided to all employees, from executives on down, on topics including teamwork, group dynamics, and valuing differences.

The change management team at a large electronics firm that was implementing a new order-management process spent considerable time trying to understand the objections of two executives on the senior management team whose participation was critical but who were not convinced of the benefits of process innovation. The team took one, the chief financial officer, to visit a winner of the Baldrige quality award, where he met with the firm's CFO. The other, who turned out to be concerned that he would not have a role when the new process was implemented, was notified by the CEO that were he to join the process innovation effort there would be an important role for him whatever the new design. His attitude changed rapidly. Not all managers and workers can be addressed in such individualized fashion, but for key executives the time spent is worthwhile.

Communications and commitment building.   Given that achieving organizational change is perhaps the most difficult aspect of process innovation, a concerted effort must be made to communicate throughout an initiative about the change program and to build commitment to the new design. The team that is leading change cannot wait until the design is complete to become concerned about change management. At each step of the approach, there must be consideration of how best to communicate the activities and of what steps can be taken to ensure organizational acceptance. Communication and commitment building must occur at all levels and for all types of audiences. Certainly senior management must exhibit positive attitudes toward the initiative and be receptive to the new design, but even the lowest-level employees whose work is affected by the changes must be prepared to alter their behavior. Customers should also be the focus of efforts to build adherence to the new process.

Although the details of communication and commitment-building programs will vary with the organization and processes involved, regular communication must be established between the executive and process innovation teams and those who will be affected by the new process. Sensitive issues, such as the level and type of personnel reductions to result from the initiative, must be addressed honestly and openly. The selection of teams should be done in a way that maximizes acceptance of change by the rest of the organization. Prototyping is another vehicle for communicating and building commitment. The entire change effort should be viewed as a public relations campaign—akin to selling a product or a political candidate.

These goals are so important that they should be specifically assigned to members of the process innovation team. The change management teams established by some firms are an obvious locus for this set of responsibilities.

Equally important to change communications is the creation

and maintenance of a sustained sense of meaning in the process innovation effort. People want to be part of something meaningful, greater than themselves.[10] A vision that galvanizes an organization to action will include an emotional call to join in the taking of the moral high ground. Visions almost always begin with calls for "Excellence" and "Quality" in an organization's products and processes, but this appeal to emotions must be combined with a bold outline of what excellence will look and feel like and how it will be measured.

Communication to the organization as a whole needs to start well before implementation begins. Once the business unit head has built commitment within the executive ranks, he or she needs to begin to send signals to and energize the rest of the business unit. The sponsor must use public events, including speeches and meetings, and other communications media to prepare and rally the troops for the changes to come. All employees need to understand the competitive threats and the vision that is taking shape, and they need reassurance about the organization's ability to improve its processes and business results successfully. The announcement of process owners and teams is an obvious example of an occasion that should be exploited by the sponsor to get these messages out.

As process teams make their initial recommendations, the level of communication needs to increase and to be supplemented with structured participation by representatives from the middle and lower ranks of the organization in prototyping and piloting activities. The activities need to be publicized, and participants need to share their experiences and the results with their constituencies and others in the organization.

The End Point Model gave DSM just such a strategic and inspirational vision, a vision that motivated the management team and fired up a sleepy manufacturing organization. The emotional appeal of the vision derived from the group manager's clarion call to join in a "Journey to Manufacturing Excellence." "Excellence" was to be measured in terms of reducing manufacturing cycle time and time to market by 50%. Subobjectives were also defined. The vision's intellectual appeal was captured in a picture that conveyed the future state of DSM. The End Point Model depicted an integrated enterprise; it portrayed people and processes within DSM and customers and suppliers' organizations "virtually" integrated by information technology.

Measures.   A successful change vision also includes the setting of clear strategic goals. Establishing measures of strategic success and setting parameters for these measures is part of the conceptualization of the vision and an important means of communicating how the organization of the future will differ from the organization of today. It is important that statements of objectives be specific and measures operational. For those who must make it happen, stating that a division will reduce the cycle time for introducing new products by 50% by 1993 is much more meaningful and helpful than stating that the division will increase revenue by 15%.

Sometimes, existing measures are left intact, but stretch objectives—dramatic increases or decreases in performance parameters—convey the outlines of a radically different organization. For example, Rockwell's telecommunications division sets targets of 50% improvement for its process improvement initiatives. When these goals have been achieved, it targets further 50% improvements. Targeting 50% reduction in the cost of securing and correctly filling a customer order is an example of a stretch objective related to a traditional measure—cost. What is not traditional about this statement is its process-oriented focus.

A vision can also be communicated through a shift in the focus of existing, or through newly created, measures. New operating visions have turned some traditional measures on their ear. Until recently, most companies set goals and measured shipping and service delivery performance on the basis of a ship date committed to very late in the order/delivery cycle. Leading companies that have switched from producer time to customer time are measuring themselves on their ability to meet a delivery date requested by the customer very early in the order/delivery cycle.

Within DSM, "excellence" took on some new and rather specific meanings. For example, as part of the planning process the team rethought and redefined the way it measured success (new and old criteria are identified in Figure 9-4).

As we noted, MRP II Class A certification was one of the first milestones of the 1986 plan. It was also the basis for measuring DSM's progress. MRP II was a plant-level project; accountability was established, starting with the plant manager, for 13 measures, progress was tracked quarterly, and the results were posted prominently in each plant. Figure 9-5 presents a representation of the MRP II Report Card posted in the Augusta plant's cafeteria.

Figure 9-4   New and Old Success Criteria for DSM

Old	New
Supply/Demand	Demand/Supply
# of Suppliers	# of Alliances
Partial Orders	Complete Orders
Delivery to Commit	Delivery to Request
Cost	Margin
Quality at Dock	Quality at Customer
% Build-to-Stock	% Build-to-Order

Figure 9-5   Augusta Plant Quarterly Report Card

Augusta, Maine: Quarterly MRP II Performance Report

	Q1	Q2	Q3	Q4	Q5	Q6	Q7
Top Management Planning							
Business Plan	0	0	100	100	100	96	92
Sales Plan	48	0	74	84	81	82	73
Production Plan	0	0	96	88	85	72	76
Operational Planning							
Master Schedule	10	22	72	65	41	88	100
Material Requirements	0	91	72	69	96	91	97
Capacity Planning	0	0	100	80	75	95	100
Database							
Bills of Material	96	97	86	100	96	100	100
Inventory Records	85	93	90	90	81	87	94
Routing	0	0	0	97	100	100	100
Operational Execution							
Material Acquisition	0	53	26	40	52	84	80
Shop Floor Control	23	19	66	54	52	97	99
Schedule Performance	99	99	99	99	97	96	97
Order Administration	63	54	96	89	93	97	98
Average	33.3	46	75	81	81	91	93
Class	D	D	C	B	B	A	A

Post-implementation interventions. However successful change efforts might be, organizations cannot remain in radical change mode forever. At the appropriate time, the behavioral and operational changes must be solidified. Two "hard" management interventions are needed to maximize gains and prevent back-sliding. The first is replacing resisters and/or individuals who have failed to adapt to the new environment, the second, making permanent changes in the organization's formal control mechanisms. Timing of these powerful interventions is critical; to achieve maximum benefit from a process innovation initiative, these interventions must occur neither too soon nor too late.

Leaders of successful process change replace resisters and those who cannot adapt to the new environment only after providing education, training, and coaching and allowing them ample time to adapt. Key executives and managers tend to be replaced during the planning phase, others during the implementation phase. Whenever replacement occurs, it sends a strong message to those who remain in the organization. But unless the replacement is perceived to be fair, there will be negative consequences.

The other hard intervention involves shifts in the mode and locus of control. These shifts must be away from the personal power and influence of the initial sponsor to the managers who now own the change, and from ad hoc structures such as process innovation teams to employees who will perform the processes day to day. For the changes to take root, control must be embedded in the organization's permanent structures and mechanisms.

DSM's group manager, like most successful process change leaders, used a combination of hard and soft interventions to manage anticipated resistance. On one hand he provided education, information, and frequent opportunities for direct, two-way communication with employees. The DSM team institutionalized the celebration of both task-oriented and behavioral successes in order to encourage change. To adhere to DEC's no-layoff policy and to encourage adoption and extensive use of the IT component of the plan, they made every attempt to find new opportunities for employees whose jobs were eliminated by the technology. One of the expert systems developed in the DSM initiative was an outplacement/career planning system.

But the group manager also displayed the impatience for results that is characteristic of successful change leaders, and did not hesitate to replace resisters and others whom he felt were not adapting quickly enough. However, some employees believed that he acted too swiftly and capriciously in certain instances. Valuable cautionary advice was discarded and resistance driven underground.

Reward and compensation systems are key means to solidifying the behavioral and operational changes that result from process innovation. Other control systems and structures need to be modified as well. Permanent operating teams need to be defined and established, and process management responsibility and new reporting relationships formalized; budgeting structures need to be modified, and new hiring criteria and job descriptions created.

Poor timing of either set of interventions—people replacement or establishment of formal control systems and structures—will yield suboptimal results. Replacing key individuals before they have been given adequate opportunity to grow and adapt will lead to demoralization and passive or suppressed resistance; waiting too long will delay progress and benefits; not replacing resisters at all will increase and institutionalize resistance because management will not be perceived as being serious.

But imposing formal changes in controls too soon, before the organization understands the need for change or the benefits of new ways of operating, will be demotivating and increase the likelihood of resistance. Employees in one firm we studied, who still had not yet broken out of their functional mind-sets, devised ways of beating a new, and perhaps not ideally designed, measurement system that had been imposed too soon in the process. In another company we studied, failure to change resource allocation policies and practices, budgeting structures, and performance measurement systems prevented some of the approaches implemented in a very successful process innovation activity from taking hold permanently.

Short-term costs.   Radical change usually incurs short-term costs, as well as benefits. DSM's rapid, measurable progress toward its goals was exploited by its leaders to create a highly charged environment in which employees at all levels made tremendous professional and personal strides and job satisfaction improved markedly. But DSM's measurable progress was accompanied by a short-term dip in one key measure during the first year—the group manager endured some tough reviews with senior management during this period as he defended the plan and held out the promise of the turnaround that ultimately materialized.

For some in DSM, the rapid and radical change, coupled with demands to maintain normal production in a rapidly expanding business, resulted in a high degree of emotional stress, verging on burnout. Inherent in the plan and the name for vision—the End Point Model—was the concept of a future state that, once achieved, would constitute the pinnacle of manufacturing excellence. But successful implementation of the plan created new insights that suggested still other ways of working. In other words, the end point was not an end after all. The new vision that evolved was of a learning organization that continuously adapts its behaviors and goals based on anticipated and unanticipated results of the implementation of the existing plan. One of the first changes occasioned by this realization was the creation of teams comprised of workers rather than executives, to review and recommend modifications to specific programs.

SUMMARY

The strategic and financial results reported in Figure 9-3 speak for themselves. In 1987, according to plan, DSM plants became customer reference sites within DEC. DSM was no longer the sleepy manufacturing organization the group manager and his team inherited in 1985. Shop floor employees conducted tours for visitors who came to see what manufacturing excellence meant. The eyes of the people who planned and implemented the End Point Model still light up when they recall and talk about this period, which had such a profound effect not only on DSM, but on the personal and professional lives of those involved. Concluding a successful process innovation program can have just this kind of benefit for all involved.

The organizational change that arises from process innovation is deep and complex. Clearly, it is the most difficult aspect of process innovation. Yet managers should not be intimidated by the variety of factors to consider and the many skills required. Most good managers have an intuitive sense for managing the required organizational changes. Following the advice in this chapter can greatly augment such managers' natural abilities.

In the next chapter, a very different type of enabler is discussed. Information technology can help to implement new process designs as well as enable them. Though IT is not normally as significant a factor in implementing processes as the management of organizational change, it can play an important role in reducing the time for a process innovation initiative and increasing the likelihood of its success.

Notes

[1]David A. Garvin and Janet L. Simpson, "Digital Equipment Corporation: The End Point Model (A)," 9-688-059. Boston: Harvard Business School, 1988.

[2]David A. Nadler, "Organizational Frame Bending: Types of Change in the Complex Organization," in Ralph H. Kilmann and Teresa J. Covin, eds., *Corporate Transformation: Revitalizing Organizations for a Competitive World* (San Francisco: Jossey-Bass, 1987); also Daryl Conner, ODR Corporation, Atlanta, Ga.

[3]David A. Nadler and Michael L. Tushman, "Beyond the Charismatic Leader: Leadership and Organizational Change," *California Management Review* 32:2 (Winter 1990): 77–97.

[4]Daryl Conner, ODR Corporation, Atlanta, Ga.

[5]Michael Beer, Russell Eisenstat, and Bert Spector, *The Critical Path to Corporate Renewal* (Boston: Harvard Business School Press, 1990): 184.

[6]Deborah G. Ancona and David A. Nadler, "Top Hats and Executive Tales: Designing the Senior Team," *Sloan Management Review* (Fall 1989): 20.

[7]Ibid., 24–25.

[8]Nadler, "Organizational Frame Bending," 74–76.

[9]Beer, Eisenstat, and Spector, *The Critical Path to Corporate Renewal*, 5.

[10]See Andrew Campbell, Marion Devine, and David Young, *A Sense of Mission* (London: Economist Books and Hutchinson, 1990); see also Beer, Eisenstat, and Spector, *The Critical Path to Corporate Renewal*, 189.

Chapter 10

Implementing Process Innovation with Information Technology

Earlier, we explored ways information technology can enable innovation in business processes. Here, we examine the potential of IT as it relates, not to creating a new process design, but to the successful implementation of an already-designed process. Implementation activities make a process design real; IT not only enables better designs, but also helps turn them into work behavior and business benefits.

It is important to keep the implementation role of IT in perspective. This chapter is one of fourteen; the proportion of attention we accord IT-based implementation is roughly equivalent to its level of importance in process innovation undertakings. We emphasize this because some firms view mastery of IT-based tools for process modeling and analysis as the primary factor in process innovation success. Such firms typically create some very sophisticated models of their existing ways of doing business, but achieve little or no business change. IT is an effective implementation vehicle for process innovation, but only when coupled with the approach, enablers, and other implementation factors described in earlier chapters.

Much of this chapter is structured around the role IT can play in facilitating activities associated with the five steps of process innovation (repeated in Figure 10-1). We discuss the prerequisites and pitfalls associated with specific opportunities to employ IT in support of process innovation objectives and examine the ways in which IT can facilitate the implementation of business processes, building, as it does so, a substantial list of enabling technologies. Finally, we integrate our activity-related observations and highlight key challenges for success.

The five major steps in process innovation enumerated in the

Figure 10-1   An Approach to Process Change

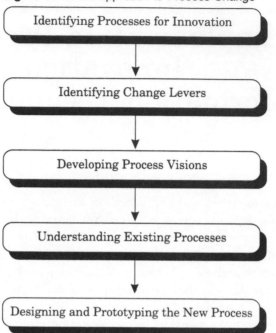

exhibit yield 10 key activities, some implicit and some explicit, in which IT can play an important facilitation role:

- identifying and selecting processes for redesign;
- identifying enablers for new process design;
- defining business strategy and process vision;
- understanding the structure and flow of the current process;
- measuring the performance of the current process;
- designing the new process;
- prototyping the new process;
- implementing and operationalizing the process and associated systems;
- communicating ongoing results of the effort; and
- building commitment toward the solution at each step.

We discuss each of these activities in turn.

IDENTIFYING AND SELECTING PROCESSES FOR REDESIGN

Identifying and selecting target processes involves information gathering and analysis. Two kinds of information need to be gathered: information about the performance and structure of candidate processes, and information about the readiness of the organization to support the redesign of those processes. To date, opportunities for IT to aid in such information gathering have been automational, as in the graphical portrayal of process flows or generation and processing of organizational surveys.

IT aids to analysis—such as systems for modeling and simulating process flows, analyzing survey and interview data, and structuring the process of evaluating and ordering a set of contending processes—can greatly increase the breadth and depth of selection analysis. They are limited primarily by the effectiveness of the people who identify and evaluate the criteria for selection. Often, these criteria are necessarily based on heavily subjective executive perspectives such as the degree to which improvement in a certain process will enhance competitive position or the extent to which there exists in the executive ranks the "critical mass" needed to support an innovation effort for a particular process.

IDENTIFYING ENABLERS FOR NEW PROCESS DESIGN

Change lever analysis, as we have pointed out, relies on both knowledge and creative thinking about how IT and innovative organizational/human resource approaches might be applied to a process under analysis. The requisite knowledge is of two kinds: (1) knowledge about the latest state and likely future capabilities of key technologies or human enablers of change, and (2) knowledge about the ways these change enablers have been or could be applied to the target process.

Databases that offer relevant knowledge of one or the other kind are available from a number of sources, and expert software and consulting services can be employed to key into the knowledge they provide, based on the parameters of a company's specific situation.[1] Such databases, which are often proprietary to specific strategy or process consulting firms and focus on IT more than on other change enablers, offer two advantages: (1) they can deliver a breadth of source data not easily matched by a manual research/interview process, and (2) they provide substantial results quickly, enabling the innovation team to focus its resources on gap filling and concentrate on the creative side of IT enablement—examining, validating, enhancing, and augmenting key results in brainstorming sessions.

When a set of technologies and potential applications has been identified for consideration, graphical organization and portrayal of the information become helpful. Useful technologies here range from desktop graphics to real-time simulations and demonstrations.

DEFINING THE BUSINESS STRATEGY AND PROCESS VISION

Developing a worthwhile business strategy and process vision relies on (1) a clear understanding of organizational strengths and weaknesses, coupled with an understanding of market structure and opportunity, and (2) knowledge about innovative activities undertaken by competitors and other organizations. IT can play a role both before and during a formal process innovation exercise.

Prior to innovation, a company's existing information systems are a major source of information regarding current capability and performance. If a firm has made a successful investment in executive information system technology, IT is likely to be playing both informational and disintermediation roles—getting information directly to senior management without the delay and possible distortion inherent in having it relayed by subordinates.

Executive information systems are valuable for providing management with a continuing understanding of business performance and the ways it depends on internal and market factors. A major pitfall of these systems is that they are usually rooted firmly in a company's current strategy and assumptions,[2] and hence are not likely to highlight issues of strength and weakness along dimensions that the company has so far ignored.

Large-scale text storage and retrieval technologies can facilitate the implementation of systems for collecting, analyzing, and disseminating information from such varied sources as news wires, commercial databases, and in-house field reports. Products in this field are being augmented by rule-based systems that enable users to specify topics of interest in order to screen out a maximum of irrelevant information. Text search and retrieval technology are growing increasingly capable with advances in artificial intelligence technology and raw computing power.

The most important requirement for effective visioning is a climate for intellectually open, creative thinking. Information technologies aimed more directly at creative thinking have application both before and during the visioning process. Computer conferencing technology, a close relation of text search and retrieval technology, enables key individuals to initiate conference topics from their workstations or join topics initiated by others.[3] Any invited individual can read entries and contribute new thoughts to an initiated topic. Topics that work well on conferencing systems are those for which a problem or challenge can be described in an initial entry, complete with additional assumptions and constraints, and for which it is desirable to solicit responses, reactions, and other constructive input. Sample topics might include "ideas for doubling our gross margin" or "key strengths and weaknesses of our top three competitors."

Computer conferencing is most effective in corporate environments in which participants' work obliges them to sign on to their workstations many times during the day. Managers who routinely use project management systems, executive information systems, personal word processing or spreadsheet applications, and electronic mail are ideal candidates. Conferencing for such managers lends an added dimension to the information-processing environment they already use. Conferencing is much less likely to succeed in an environment in which key potential contributors have little other reason to use their workstations.

Once a formal process innovation effort is under way, the key visioning activity becomes face-to-face brainstorming sessions among the appropriate participants. There are several emerging technologies for facilitating this type of group interaction.[4] Some groupware technologies involve the use of networked workstations to let participants enter into real-time discussions or brainstorming sessions, while capturing and structuring ideas in an efficient way. Discussion topics are presented on a large screen and modified or added to as discussion progresses. High-level process or financial models can be presented and subjected to real-time group "what-if" analyses. These systems can also support group brainstorming by participants who are not all in the same location. We do not know of any process innovation efforts in which the visioning activity has been significantly supported by group technologies, though there seems to be a good fit between the problem and the technology.

However, another cautionary message is warranted. Technology should be introduced into the brainstorming setting judiciously. IT can serve as a valuable process or content aid in such sessions, but also, unfortunately, as a debilitating distraction from session goals. Brainstorming works best when the chief distractions are fresh ideas—and there are rules for managing these, as well.

UNDERSTANDING THE STRUCTURE AND FLOW OF THE CURRENT PROCESS

Graphical documentation and/or process modeling tools are useful for arriving at and documenting a common understanding of current processes. At minimum, IT can facilitate the creation of informative process flow charts. But a tool that has modeling or simulation capabilities can play more than the role of artist and publisher. It can take account of current operational variables in designing and running a model that illustrates the operation as well as structure of a target process.

A wisely chosen process modeling tool satisfies four criteria:

1. it is fast and easy to use at a high level, perhaps even as early as during visioning;

2. it is applicable to the portrayal and analysis of the new (innovated) process, enabling new and old processes to be compared in the same formats and perhaps even driven by the same set of simulation variables;

3. it provides not only a descriptive, but also an analytical, model of the process, facilitating an understanding of such factors as time, cost, and other resources consumed by the process; and

4. it supports the addition of successive levels of systems and data-oriented detail, enabling it ultimately (and seamlessly) to serve a useful purpose during the systems design and/or prototyping stages.

Although we have identified no tool that meets all of these criteria, one will undoubtedly be forthcoming. The first criterion—ease of use—is delivered by widely available drawing and drafting applications that can readily create boxes, arrows, and labels in a form attractive enough to appeal to senior management. Some simulation tools meet the first three criteria, but not the fourth. Tools that facilitate systems design and delivery under the information-engineering paradigm can deliver the fourth systems-oriented objective, but are difficult to use and do not permit collection and analysis of process performance information. They also produce unattractive graphics, which can extinguish any senior management interest in process designs.

Current information systems should be analyzed for how well they support the existing process, and may be a source of information about process structure and flow. If the existing process and the systems that support it are documented with computer-aided software engineering (CASE) tools, the innovation team can derive from the resulting models an understanding of how the process is supposed to operate.[5] If, however, a CASE-based model does not already exist, it is probably not worth the effort to create one. A high-level model without detailed data flows is usually sufficient.

Adding further options for current-state systems analysis is the emergence of a growing range of "software reengineering" tools and services.[6] Conceived as a means of improving the technical state and internal documentation of current systems, reengineering technology has promise for converting a 20-year-old, poorly structured, heavily patched system into almost as attractive a platform for major new enhancements as a system that was completed only a month ago using the latest tools and techniques. The logic and functionality of the old system is modeled and a new system generated based on that model.

Reengineering technology can be an attractive and valuable expedient for capturing and documenting the essence of current systems, but innovation teams must take care to manage the scope and extent of the task. Applying reengineering technology to systems and databases that do not pertain directly to the process under analysis can cancel out the time savings such tools afford. Statements like, "This tool makes it easier for us to flesh out more of our enterprise systems and data models, which we should be doing anyway," can be very dangerous. It is also dangerous to carry software reengineering activities beyond the high-level documentation stage and into full implementation before the innovation team has ascertained that there will be a place for the current-state design in the context of the new business process. This pitfall is signaled by a remark such as, "Why not fix the system if we've come this far and have such a capable tool? We'll be able to enhance the system to provide the needed new functionality."

DESIGNING THE NEW PROCESS

Only when the process innovation team has established a high-level vision of the new process, set measurement criteria, gained a perspective on the current process, and identified the key IT enablement opportunities should process design proceed in earnest. At this point, the focus of the effort, and of IT use, is on modeling and analysis of the new process.

The design process begins simply, by starting to build on the high-level process concept that developed during the visioning stage. Depending on the tool chosen to analyze and portray the process up to this point (the visioning stage, and perhaps aspects of the current process analysis), the shift toward detailed process design can be seamless.

Ideally, a tool chosen for new process design would be capable of:

- graphically portraying the process steps;
- depicting the flow of materials and information between steps;
- accepting and portraying flow rate, resource and time consumption, and capacity and/or trigger information for each process step;
- rolling up or exploding the steps of the process in a hierarchical fashion to accommodate varying levels of detail;
- presenting a highly interactive and preferably graphical user interface;
- running live simulations and producing real-time graphical output;
- identifying key bottlenecks and constraints in the process; and

- linking to data and procedure modeling aspects of the CASE tool set to be used in IT-based system design.

Again, we know of no tool currently available that provides all these capabilities. Consequently, firms should be skeptical of vendors or consultants who approach new process design from a primarily tool-based perspective. This is not to say that tools have no use in facilitating new process design. American Airlines is experimenting with the use of expert systems to design processes and select among alternative designs, and other firms, primarily software vendors, are experimenting with process design tools that would assemble designs and software to support them from simple building blocks of common business transactions.[7] We know of no successful process innovation initiative that has used such tools, however.

A description of the IT-based systems needed to support a new process is an important aspect of the process design. The degree of accuracy and completeness of such systems requirements will have a direct impact on the duration of the implementation effort and the appropriateness of the systems ultimately delivered. Process innovation teams should be careful to first establish the design parameters of the new process and then think about systems to help implement the design. If the two objectives can be jointly accomplished with a single tool, so much the better. In any case, process design work should provide a solid lead-in to formal systems design work.

When consensus has been reached on the high-level flow and attributes of a process design, ancillary materials should be gathered, namely:

- the business/process vision developed earlier, coupled with a description of how the new process relates to the physical organization of the enterprise;

- a description of the information systems requirements (functionality) called for by the new business process; and

- a high-level model portraying the flow and use of data within and around the process.

Information engineering-based CASE tools and methods—described by James Martin as "an interlocking set of formal techniques in which enterprise models, data models, and process models are built up in a comprehensive knowledge base and are used to create and maintain data processing systems"[8]—have the capacity to support the development of the latter two. Having achieved mainstream status in the systems development world, information engineering is widely recognized to benefit systems development in several ways.

1. It brings engineering rigor to a discipline renowned for producing an abundance of errors and oversights in deliverables.

2. It emphasizes systematic contact with users over the development lifecycle to ensure an accurate translation of user requirements into an operational system.

3. It leverages the automational, integrative, and analytical enablement aspects of current CASE technology, including graphical modeling, data analysis and sharing, process analysis, and modularity.

Information engineering is currently regarded by most experts as the preferred method for rigorously defining and developing the systems needed to support new process designs. It takes processes as a fundamental unit of analysis and includes a step for redesigning them. But information engineering as performed by most organizations has some shortcomings in a process innovation context.

The information-engineering approach includes four phases: systems planning, business area analysis, systems design, and construction. Process redesign is supposed to take place during the business area analysis phase, which is too late to initiate process innovation. Only in the planning phase can we find the strategic context and high-level management participation needed for high-level process objectives and attributes to be defined. If process innovation is not the focus of the planning phase, it is unlikely to happen at all.

Even if one were to attempt to redesign processes in an information engineering context, the processes defined under the method are too narrow. Information engineering, as we noted earlier, defines processes as smaller units than we have been discussing. In fact, it places them fully within functions. Finally, although it includes a step for describing old and new processes, information engineering offers no information about how to effect process improvement or innovation or about the role of IT in doing so. In short, information engineering is not a methodology for process innovation; to be used as such, it must be modified substantially.[9]

Moreover, issues of scope and focus in information engineering require careful consideration. Under ideal circumstances, information engineering calls for a data model of the entire enterprise or, at the very least, a large segment of the business. Although such a model can pay dividends in maximizing an organization's understanding and efficient use of data, it has three drawbacks in a process innovation context. One, it takes time, often more than a year, to develop an in-depth model of an enterprise. This is usually much too long for enterprise modeling to be a step in a process innovation initiative. Two, the resulting model will be rooted largely in the present and its very mass and level of intellectual investment will bias the organization toward evolutionary rather than revolutionary process thinking. Finally, the level of detail produced by an enterprise modeling effort guarantees that only relatively low-level workers will want to be involved and senior management interest will be low.

For radical process innovation to succeed, it is necessary to restrict the scope of effort to the immediate process and its direct interfaces. It may also be necessary to confine detailed data modeling to subsets of the overall data environment. Even outside a process innovation initiative, many firms have difficulty reconciling the costs and benefits of enterprisewide data modeling.[10]

In organizations that have already developed substantial enterprise models and systems plans, process innovation teams must exercise caution in relating the process analysis to existing specifications. Often, the elements of relevance to the new process will span many organizational or functional areas in the original model, a consequence of the likelihood of contemporary business processes to cross organizational and functional lines.

PROTOTYPING THE NEW PROCESS

The prototyping of new processes usually overlaps to some degree the design activity and can begin as soon as the process design begins to take shape. Prototyping maximizes participant exposure to physical aspects of a process while it is being designed. Prototyping enables process, systems, and organizational design possibilities to be presented, approved, and/or modified by users before the organization has invested much effort in detailed design and implementation. Thus, process design and prototyping activities are iterative.

IT can speed and increase the productivity of prototyping. Looking first at prototyping of process flows, real-time simulation tools are of value. In the context of prototyping, humans may play mock roles at one or more stages of the process (typically, at minimum, at the first and last steps). Simulation tools can demonstrate the remaining aspects of the process. To achieve seamless integration with the process design activities, the analytical tool chosen for process design should include the necessary simulation capabilities.

IT can provide leverage in prototyping efforts related to the information systems associated with a new process in two ways. One is via the ability of many advanced CASE systems to enable a design team to select and bring key aspects of the systems design to a relatively advanced state. This design includes sufficient information to permit the CASE tool set to generate (inefficient but) working program code to implement the system. The other is by employing programming environments termed "very high-level languages" to facilitate rapid development of application functionality. This approach is most effective when the aspects of a system that require prototyping are segmentable into small, independent, and well-defined units. At minimum, such units function as storyboards that demonstrate successive screens and/or input/output actions of an intended system. But they may also be completed components that will be in some way integral to the pilot and/or delivered system. In a prototyping context, all of the tools listed in Figure 10-2 could qualify as very high-level languages.

The last of these, code generation, is a specific example of very high-level language programming that stands side by side with a range of additional options that may be as (or even more) expedient for a particular prototyping need.

In considering which IT option for prototyping will integrate most seamlessly with subsequent development of production versions of future information systems, design team members should not lose sight of the rationale for prototyping: to give users hands-on experience with key aspects of a system at as early a point as possible in the design process. IT choices that maximize the benefit to a specific process innovation effort do not always integrate well with production development tools.

Figure 10-2 Very High-Level Language Programming Environments Useful in Prototyping Activities

Fourth-generation languages

Object-oriented languages coupled with class libraries

Conventional languages coupled with subroutine libraries

Programmable database packages

Programmable spreadsheet packages

Programmable hypermedia packages

Storyboarding packages

Code-generating CASE tools

IMPLEMENTING THE PROCESS AND ASSOCIATED SYSTEMS

Project management tools can benefit process implementation in two ways. One, they can aid in identifying, structuring, and estimating design, construction, and implementation activities. With individual work steps in a major innovation project likely to number in the hundreds, project management tools are an important vehicle for saving time and enhancing project plan quality. Two, project management tools facilitate on-the-fly analysis of ongoing project management and scheduling developments. Used properly, a project management tool can greatly enhance a manager's ability to control a large project and address contingencies in a timely and effective manner.

Process implementation can also be abetted by machine encoding of the project approach. A number of the systems development lifecycle (SDLC) methodology products available today incorporate work plans in electronic form, often already loaded into an integrated project management tool. Sometime soon, SDLC project management systems may be able to manage the launch of CASE tools for a systems developer, loading the information appropriate to the particular work step and guiding the developer through the appropriate use of the tool and associated knowledge bases and back to the work plan, ready to select and attack the next work step.

Rapid systems development approaches that support changes in the sequence of systems development work are highly relevant to process innovation implementation. Made possible by advances in information engineering-based CASE tools (enhanced code-generation capabilities, for example) and prototyping options, these approaches can enable implementation activities to proceed concurrently with design, substantially conserving both calendar and applied time.[11] These benefits are both welcome and essential, as timeliness is an end as well as a means in process innovation efforts.

Information engineering-based CASE tools are as relevant during implementation as during design. And, as during design, the scope of projects that employ these tools should be managed so as to coincide with the objectives of the business/process vision established at the outset, and rapid development techniques should be applied wherever possible in place of the more traditional "model, then design, then code" approach.

An emerging (in commercial environments) enabler of process implementation is object-oriented systems design and programming. Well understood in academic and research circles, object-oriented techniques enable programs to be shared by higher-level programs that need their functionality rather than be reimplemented over and over again. The reduced redundancy pays dividends in reduced development time and improved quality.[12]

A secondary benefit of object-oriented design is that it forces developers to structure applications to run efficiently on parallel processing architectures. This is particularly important for processing-intensive applications such as image processing or large-scale database searching, and computation. The processing power of single-processor computers is growing at about 30% per year, compared with five- to tenfold for massively parallel computers. To the extent that an application can be broken up to run in parallel pieces, it can benefit from the incredible throughput potential inherent in parallel architectures.

The long-term goal of firms that are beginning to experiment with the relationship between processes or business rules and systems objects is to be able to spread "modules" of data, applications functionality, and process throughout an organization.[13] Similar modules might, for example, be applied in different divisions or geographical units.

In a system designed along these lines, each object might use different types of software technology. Data objects could be managed with object-oriented database-management software, which is still in a relatively early stage of technical development. Applications logic could be stored and maintained in traditional algorithmic programming languages. Business rules could be programmed in the same technologies used for rule-based expert systems, which could facilitate their structuring and modification. It should be emphasized, however, that it will be several years before this sort of environment is practical.

A slightly more mundane version of the process-inclusive object is the application package that takes the form of a process model. Some vendors are beginning to supply large-scale applications packages in the form of process and data models, which users can tailor to their companies' own processes before generating systems that fit the desired processes exactly. We expect to see more examples of the relationship between business processes and systems functionality over the next several years.

COMMUNICATING ONGOING RESULTS OF THE EFFORT

Advances in IT have been directly responsible for many improvements in systems-development productivity and quality over the years. But nontechnical developments can also drive improvement. An example is the ever-stronger tenet that process implementers (that is, systems developers) should be in as frequent and complete communication with process users as possible. Although the notion has existed for centuries, its steadily growing centrality to implementation methodologies warrants its classification as "an advancement."[14]

Information technology's signal role in supporting ongoing communication is characterized primarily by the many work-step deliverables of information engineering and project management methodologies and the almost continuous communication that is characteristic of prototyping work. Perhaps the most well-developed example of structured communications between designer and user is the "joint application design," in which a system's creators and users engage in a series of structured, jointly participative meetings.[15] Although there exists no structured equivalent of this approach for process innovation, some of the same principles may be applied to communicating about process innovation.

Any strategy that involves drowning team members and users

in piles of paper warrants judicious application. Frequent deliverables are essential, but care must be taken to ensure the usefulness of their content. The goal is to be aware of exactly which issues need to be communicated at a given time and to ensure that these are not lost in irrelevant communications.

BUILDING COMMITMENT TOWARD THE SOLUTION AT EACH STEP

Communication at each stage is a prerequisite to building the commitment that will ultimately result in an organization adopting a new business process. In addition to enabling the generation and communication of deliverables, IT plays several indirect (and admittedly relatively minor) roles in commitment building. One, project management tools, with their tracking enablement, represent a valuable aid to the innovation team as it works to continually manage expectations against commitments against deliverables. Two, although overanalysis in an information engineering context is dangerous, the various activities and deliverables of information engineering have a ritual value; they can, by their very nature, serve to strengthen relationships between the designers and targets of new processes. The trade-off between ritual and rapid benefit should represent one of the factors that drives scope management decisions. And three, communications technologies such as telephones, electronic mail, voice mail, conferencing, and so forth play a crucial role in overcoming geographical barriers to communication. Whether a particular communication includes or refers to a formal deliverable in the innovation process, the odds are that it represents an incremental improvement in relations among key process innovation constituencies.

The preceding analysis has focused on ways IT-based tools enable activities in the various process-innovation steps rather than on how they might serve to automate analytical techniques. IT-based tools and techniques are important only to the extent that they represent current and effective approaches to achieving process innovation objectives. They lose their relevance when superseded by new tools or techniques or when they are no longer applicable to a target process.

An overt focus on tools and techniques—IT or otherwise—

can stand in the way of achieving process change objectives. Particularly in the case of tools and techniques that are complex and unfamiliar, ascending the learning curve may shift the focus away from business change. It can be tempting to concentrate on mastering a technology rather than attacking the much more difficult issues of organizational inertia and change.

THE RELATIONSHIP BETWEEN PROCESSES AND INFORMATION SYSTEMS

The ultimate goal of a business process innovation effort is to deliver to the enterprise a new process that achieves one or more revolutionary performance objectives. It is not to design, code, and test an information system. The objectives of business process innovation fall into three groups, which tend to lead into one another. Strategic objectives lead to process-related objectives, which lead to information systems-related objectives. In short, information systems objectives are only a portion of the process innovation challenge.

It is important to apply information systems-oriented approaches such as information engineering with all these objectives in mind, and to recognize that the tools and techniques for achieving process objectives are distinct in use from those for working toward systems development objectives. One way to inculcate this mind-set is to give the strategic and process domains a head start in a project and move through most of the visioning work before engaging in any information systems design or development.

The substantial overlap in terminology between the respective tools and techniques should not be equated with integration, which is in some cases illusory. For example, "process modeling" in traditional information engineering works well in software systems analysis, but does not accomplish many of the goals necessary for "process modeling" in process design. Similarly (and in this case, within a single objective group), process modeling is done differently in support of current operations modeling than in new process design, and the two analyses may never specifically relate to each other.

Although the techniques are distinct among strategic, process, and information systems-related activities, coordination among the three domains is essential. Applications requirements and data structures that emerge from information systems-related activities should fit well with the corresponding processes, and, in the spirit described for prototyping, systems implementation activities should coordinate closely with corresponding process implementation efforts.

SUMMARY

By now it should be clear that IT can help in many ways and at many points in the implementation of process innovation. We have addressed the many pitfalls associated with the aggressive use of advanced tools and approaches, but there are also dangers associated with failure to pursue these opportunities. Three are paramount.

- The success of process innovation often lies in the speed with which it can be accomplished; failure to employ IT enablers can substantially compromise design and implementation speeds.

- A new business process is less likely to succeed if the initial implementation is of low quality; advanced systems analysis and project management techniques enhance quality.

- Failure to recognize and act upon IT's potential to benefit process innovation implementation may be an indicator of insufficient management awareness of process innovation as it relates to the key processes of the business—an indicator, in other words, of impending failure.

Our examination of the activities associated with business process innovation included discussion of the technologies (listed in Figure 10-3) that can play a role in the implementation of process innovation. Firms that are serious about process innovation should begin to master these technologies and create an infrastructure of tools and skills that can be applied to multiple process initiatives.

Although these advanced tools and capabilities are clearly relevant to the design and delivery of information systems, process innovation teams are advised to ensure that their approaches to IT implementation leave room for adequate handling of the other key aspects of process innovation—business and process analysis, design, implementation, and operationalization.

Figure 10-3   IT Enablers of Process Innovation Implementation

Computer-aided software engineering (CASE)
Code generation
Conferencing
Conventional programming
Current applications
Data gathering and analysis tools
Decision analysis software
Desktop graphics tools
Executive information systems (EIS)
Fourth-generation languages
General communications technologies
Group decision-support systems (GDSS)
Hypermedia
Idea generation tools
Information engineering
Object-oriented programming
PC-based prototyping tools
Process modeling tools
Programmable databases and spreadsheets
Project management tools
Prototyping
Rapid systems development techniques
Simulation
Storyboarding
Strategic application databases (generic and case-based)
Systems reengineering products
Technology trend databases
Very high-level languages

Notes

[1]One such tool is described in Cornelius Sullivan, "Reasoning by Analogy—A Tool for Business Planning," *Sloan Management Review* (Spring 1988): 55–60.

[2]For a discussion of how EIS relate to strategy and management goals, see John C. Henderson, John F. Rockart, and John G. Sifonis, "A Planning Methodology for Integrating Management Support Systems," in John F. Rockart and Christine V. Bullen, eds., *The Rise of Managerial Computing* (Homewood, Ill.: Dow Jones-Irwin, 1986): 257–282.

[3]For an example of how conferencing systems work in practice, see Lynda M. Applegate and H. Smith, "IBM Computer Conferencing," 9-188-039. Boston: Harvard Business School, 1990.

[4]For a broad discussion of group technologies, see Robert Johansen, *Groupware: Computer Support for Business Teams* (New York: Free Press, 1988).

[5]Business modeling with CASE tools is described in Robert L. Katz, "Business/ Enterprise Modeling," *IBM Systems Journal* 29:4 (1990): 509–525.

[6]For a discussion of this technology, see "Reengineering Software," *InformationWeek* (May 27, 1991): 62.

[7]The work of Fernando Flores in this area is described in Peter G. Keen, *Shaping the Future: Business Design through Information Technology* (Boston: Harvard Business School Press, 1991).

[8]James Martin and Clive Finkelstein, *Information Engineering,* vols. 1 and 2 (Carrforth, Lancashire: Savant Institute, 1981).

[9]Ernst & Young is in the process, for example, of modifying its information engineering methodology to accommodate process innovation.

[10]Dale L. Goodhue et al., "Strategic Data Planning: Lessons from the Field," working paper CISR 215, MIT Sloan School of Management, Center for Information Systems Research, October 1990; Dale L. Goodhue, Judith A. Quillard, and John F. Rockart, "Managing the Data Resource: A Contingency Perspective," CISR 150, MIT Sloan School of Management, Center for Information Systems Research, January 1987.

[11]For a discussion of rapid systems development approaches, see Jack B. Rochester, "Improving Application Development Productivity," *I/S Analyzer* 28:3 (March 1990): 1–12; and Robert Moran, "RADical Application Building," *ComputerWorld* 24:4 (January 22, 1990): 27, 32.

[12]Ronald J. Norman, "Object-Oriented Systems Design: A Progressive Expansion of OOA," *Journal of Systems Management* 42:8 (August 1991): 13–16.

[13]See, for example, Robert P. Dunham, "Business Process Management Technology—Software Development for Customer Satisfaction," Action Technologies, 1990.

[14]For a historical perspective on user relations, see Andrew L. Friedman, *Computer Systems Development: History, Organization, and Implementation* (New York: John Wiley, 1989): Part 3.

[15]Darrell S. Corbin, "Team Requirements Definition: Looking for a Mouse and Finding an Elephant," *Journal of Systems Management* 42:5 (May 1991): 28–30; Allen Gill, "Setting Up Your Own Group Design Session," *Datamation* 33:22 (November 15, 1987): 88–92; Don Leavitt, "Team Techniques in System Development," *Datamation* 33:22 (November 15, 1987): 76–86; and "Using IBM's Joint Application Design (JAD)," *EDP Analyzer* 24:6 (June 1986): 13–14.

Part III

Innovation Strategies for Typical Process Types

The previous sections of this book have addressed the "process" of process innovation, including the approach and implementation. It is time to focus on the content of process change. The following section describes the strategies that companies are employing to redesign key processes. Each chapter deals with a different type of process. As in earlier chapters, we discuss both technological and human enablers of these innovations.

Just as it is difficult to select processes for redesign, it is difficult to isolate processes to describe. Some of the same issues that influence the development of a process architecture for a particular firm are relevant to the structure of the following chapters. Should manufacturing be treated separately, or as a part of order management? Is research a process or a subprocess? No architecture of processes is without its compromises, and the choice of topics in these chapters is also driven partially by expediency.

The discussions of process strategies and enablers that follow are an overview. Experts in any particular process domain should look elsewhere for detailed insights. However, the innovation strategies adopted by other firms that are described below can be a starting point for the development of a process vision. More detailed research and benchmarking should be undertaken in order to yield a vision more relevant to a firm's specific environment.

Chapter 11

Product and Service Development and Delivery Processes

Because nothing is more critical to a firm's competitive success than its ability to develop new products and services and deliver them to customers, product/service development and delivery processes are likely candidates for innovation in virtually any company. Yet few companies have adopted a process view of these activities, applied innovative thinking to the processes, and employed information technology or human resources to enable radical change. This chapter, a high-level guide to IT-enabled process innovation in product/service development and delivery, examines both "make and sell" businesses and service companies that rely on research, design, engineering, manufacturing, and logistics processes.

Edward Roberts observes in a review of the management of product innovation that "the management of technological innovation is complex, involving the effective integration of people, organizational processes, and plans."[1] Because a great many processes are linked, innovation both in product development and in delivery often consists of improving interfaces between processes and subprocesses. Many firms that excel in the manufacture of complex products but falter in delivery do so, for example, because of slow or inaccurate internal financial systems (pricing, billing, and so forth). Such a firm might squeeze every minute and cent possible out of manufacturing, yet face days or weeks of delay in getting the order out of the factory.

Product development and delivery processes used to be largely back-room processes, devoid of customer contact. Today, there are no back-room processes. Competitive demands and customer insistence that process performance is as important as product quality and price have given rise to an environment in which processes must be integrated. The result is product and service development and delivery processes that are inextricably linked to customer-facing processes. The price of this linkage to companies redesigning their processes is iteration; after sequentially redesigning order management and customer service, the company must go back and adjust order management again. In our treatment of this iteration, the price is some overlap in the discussion of how different processes relate to one another. But these costs are small relative to the benefits to be derived from thinking about linkages between processes and the flow of work through an organization.

We discuss product and service development and delivery processes roughly in the order in which they occur in companies. We include examples of innovation in firms that have not adopted an explicit process orientation, but in which innovation is applicable in a process context.

RESEARCH PROCESSES

Research, being but one component of the product development process, should not be the exclusive focus of a process innovation initiative. But because it often represents the beginning of the product development process and, in some industries, is a key source of product development delays, excessive costs, and unsuccessful products, it warrants individualized attention.

It is important to view research as part of a broad, encompassing approach to product development. In fact, it is usually only a subprocess, done either before (in the case of basic research) or after (for applied research) the conceptual design of a product. See Figure 11-1 for a typical flow of a new product-development process.

Many product development delays occur because of poor coordination among research, engineering, and manufacturing. Scarce

Figure 11-1   New Product-Development Process

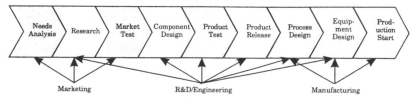

time and resources are frequently devoted to the research and engineering of products that have no place in the development pipeline. In order for unpromising products to be killed as early as possible in the product development process, and to ensure that development efforts fit a firm's product strategy, the product development process must be closely tied to the strategic-planning process.

As a group, research employees are the least likely to view their work in process terms. Yet, in our experience, the work of such employees can be readily mapped in process terms, that is, in terms of work flow, inputs, outputs, customers, time, and cost, and there are distinct advantages to doing so. But there is also an undeniable tension created when professional researchers are overmanaged. More so than those who work in other functions, researchers must discover the benefits of process management themselves.

A process view implies greater visibility and monitoring of research deliverables than is afforded by the traditional "black box" view of product research, in which money and highly educated researchers go into the box and product ideas come out. Increasing emphasis on speed in product development renders the black box approach untenable in most organizations. Time and cost savings are the paramount objectives of most companies attempting to innovate in the research process. An oft-cited study of high-technology industries reports that on-budget products that enter the market six months late earn 33% less profit over five years, whereas products 50% over budget that enter the market on time earn only 4% less profit over the same period.[2] Many firms have successfully reduced their product development cycles, but accounts and descriptions of their success seldom describe the research process (or subprocess).[3]

In one pharmaceutical firm we studied, each day's delay in submitting a drug to the FDA and receiving its approval cost the firm $1 million in revenues. While several firms are working with the FDA to develop a so-called Computer-Aided New Drug Application (CANDA) in order to help the FDA do its job faster, not all of the delays in getting drugs to market can be attributed to the FDA. A few other pharmaceutical firms are beginning to address the segments of the research process that precede FDA involvement.

Firms that attempt to innovate in the research process tend

to focus on the tracking of research projects and the matching of projects to scarce resources. (Some examples of change enablers in research processes can be found in Figure 11-2.) Such firms build research databases to accumulate project information, report results thus far, and project the timing and duration for which specialized resources will be needed. Firms can manage this information to accelerate or slow projects so as to maximize resource use and identify less-promising projects for early cancellation. Project status and historical information are very useful for presentation and discussion during "gate reviews"—formal, cross-functional meetings conducted during different stages of the development process to ensure that resources are focused on the most promising projects and that new products meet business needs.[4]

Firms in the agricultural chemical industry that must test promising chemical compounds in highly controlled field experiments are using information technology enablers to speed process in at least two ways.[5] One, scientists using laptop computers in the field are able to get data to headquarters more quickly for analysis and interpretation. But even more significantly, field computers are enabling some firms to conduct research design and analysis in the field, thereby reducing the number of failed experiments linked to local conditions. Similar benefits have been mentioned by clinical trials managers in pharmaceutical firms who have provided field computers to clinical physicians.

Although not implemented within a conscious process context, such innovations could constitute an integral part of an overall process-innovation effort. Nevertheless, the structure and measurement that a process perspective might lend would make the innovations more valuable and more likely to succeed, and we are familiar with a number of companies that are investigating innovations in research in a process context. Recent proponents

Figure 11-2   Enablers of Innovation in Research Processes

- Computer-based laboratory modeling and analysis
- Computer-based field trials and communication of results
- Tracking and project management systems
- Wide dissemination of information on project status

of new approaches to research management advocate process-oriented approaches. Roussel, Saad, and Erickson, for example, while not consciously adopting a process orientation, espouse as part of what they term "third-generation R&D"[6] a number of management practices that have a strong process flavor. Among them are the following:

- rigorous communication, throughout the organization and using a common vocabulary, about research projects and their status;

- clear and measurable project objectives;

- a "process" (their term) for setting priorities for and allocating resources among projects; and

- a multidisciplinary, output-oriented organizational approach.

Combined with the power of information technology and other enablers of innovation, these elements of process thinking can yield the levels of improvement in research processes demanded by the contemporary business environment. But they must be carefully balanced against the historical independence and autonomy of scientists and research professionals; this will ensure that the human resource element in research processes remains a key enabler of product and process innovation.

ENGINEERING AND DESIGN PROCESSES

Innovations in engineering and design share with those in research processes the primary objective of getting products to market faster. "Concurrent engineering" and "design for manufacturability" are widely advocated, if not as widely employed, means for reducing cycle time. Important though these concepts are, other objectives must not be neglected. Designing products that are valued by the marketplace and developing them at reasonable cost are also important and should be managed as process objectives. Enablers of engineering and design process objectives are identified in Figure 11-3.

Concurrent engineering entails a switch from serial to parallel process flow (the phrase Japanese firms use to describe concurrent engineering is "parallel approach"). Whatever term is used, the concept involves designing and engineering multiple

Figure 11-3   Enablers of Innovation in Engineering and Design
Processes

- Computer-aided design and physical modeling
- Integrated design databases
- Standard component databases
- Design-for-manufacturability expert systems
- Component performance history databases
- Conferencing systems across design functions and among design, manufacturing, and sales
- Cross-functional teams comprising individuals from design and manufacturing

components or aspects of a product at the same time. The success of the interaction involved in designing the product simultaneously with the manufacturing process frequently determines both the time requirements and cost of the development process.[7]

To succeed with concurrent engineering requires a high degree of organizational integration within the engineering function and between engineering and other functions. Beyond organizational integration, the primary enablers of concurrent engineering are computer-aided design and engineering (CAD/CAE) tools and cross-functional teams.

CAD/CAE workstations, software, and networks support the rapid creation and modification of two- and three-dimensional designs and facilitate the communication of designs throughout a function or around the world. The precision and simulation capabilities of these tools permit an engineer, using only high-level specifications, to design components that will interface with others, and recent advances in rapid prototyping devices make it possible to create physical prototypes almost immediately.

Not only the tools, but also rigorous construction and enforcement of engineering data models and definitions, are necessary for successful concurrent engineering. Development of a common, worldwide CAD system and common engineering data structures, for example, enabled Black & Decker's power tools division to halve the product development cycle for new power tools, from

18–24 months to 9–12 months.[8] These measures were not a luxury in an industry with aggressive Asian competition.

CAD/CAE tools have found wide acceptance in the automobile industry, both in Japan and the West. Nissan, believed to be a leader in the use of computers to integrate the design process, claims to have reduced design cycle time by 40%.

Although such technologies have reached high levels of penetration in the engineering function, it is widely known that their capabilities have not been fully utilized. While there are some technical problems, for example, in sharing designs between engineering and manufacturing, the more serious barriers to utilization involve poor integration within functions.[9] Expecting CAD systems to lead to such benefits without an explicit process change orientation is unrealistic.

Similarly, the use of cross-functional organizational teams to eliminate intra-engineering boundaries is frequently recommended in concert with CAD/CAE. However, such cross-functional teams of individuals of diverse perspectives and backgrounds are difficult to establish. Cultural differences between design and manufacturing engineers begin in engineering schools. One study of design-manufacturing integration found that even in an organization that had successfully implemented cross-functional teams with overlapping project phases, design engineers still tended to make proposals and manufacturing engineers to respond to them.[10] In another firm we studied, manufacturing and design engineers were separated both by distance (they were in separate locations, and the design engineers often traveled) and time (design engineers did their work on products months and even years before they were scheduled for production).

An example of the successful combination of organizational and technical enablers of design processes is found at Kodak. The firm made extensive use of CAD, cross-functional design teams, and an integrated design database in order to reduce development time for its new Fling 35 single-use camera by 40%. According to the manager who led the design process, organizational issues associated with this approach were much more difficult to manage than technical issues.

An extreme of teaming, the "skunk works," isolates a team comprising all the members needed to design a product and move it to production. The difficulties associated with establishing a successful skunk works are evidence of the magnitude of organizational change that process innovation entails. Lockheed's Kelly Johnson, a celebrated practitioner of the skunk works philosophy, describes these issues.

> The ability to make immediate decisions and put them into rapid effect is basic to our successful operation. Working with a limited number of especially capable and responsible people is another requirement. Reducing reports and other paperwork to a minimum, and including the entire force in the project, stage by stage, for an overall high morale are other basics. With small groups of good people you can work quickly and keep close control over every aspect of the project.[11]

These initiatives sound like key aspects of process innovation. Were they likely to employ a formalized process approach, skunk works and other types of teams might achieve higher rates of success. Skunk works organizations are sometimes viewed as a reason for lack of discipline. The absence of a process orientation may inhibit the engineering executive's understanding of the commitment and change necessary to make such teams effective.

The underlying notion in design for manufacturability, another key concept in bringing innovation to engineering and design processes, is that choices made in the product design process have significant implications for the ease and quality of manufacture. In a recent survey, senior executives in the electronics industry cited poor design for manufacturability as the most frequent cause of later-stage delays in getting new products to market.[12]

Growing attention to design for manufacturability is evidence that companies are beginning to recognize the importance of engineering the manufacturing process as well as the product. Two influential IBM and General Electric research and engineering executives put it thus:

> The design phase of the cycle of development has traditionally concentrated on the features and performance of the product rather than on the processes by which it is manufactured. We design a product first and then tackle the job of how it is to be made. Yet the eventual cost and quality of the product is inseparable from how it is made. If a product can be made easily, its costs will be low and, most probably, its quality high.[13]

Whereas concurrent engineering is primarily an intrafunctional process, design for manufacturability is clearly oriented toward eliminating functional barriers between engineering and manufacturing. To the extent that the organizational structures of manufacturing and design are factors in its success, design for manufacturability relies heavily on improved communication between these functions, which can be facilitated by cross-functional teaming.[14] But as with concurrent engineering, communication between engineering and manufacturing cannot be exclusively via ad hoc contacts within teams; it must also be structured in terms of the development of formal design standards and data models that specify preferred design and component choices. This may be difficult to do using traditional types of information systems. Xerox's development and manufacturing division has developed a prototype expert system to facilitate design for manufacturability that, when fully deployed, will help engineers evaluate alternative designs using internal Xerox information about what components already exist or could be easily sourced.[15] The system discourages the use of components that are totally new or that require extensive additional work. Coupled with efforts by engineers to reduce manufacturing costs, the system can greatly facilitate the new product-development process.

Although the effectiveness of product design processes is most frequently addressed in terms of new designs, other factors contribute to slow and inefficient processes. Reducing the number of changes in product development cycles, for example, can yield significant process benefits.

Relevant process interfaces are between engineering and sales/marketing on one hand, and manufacturing on the other. On the sales/marketing side, many firms face the problem of changing product specifications frequently to address needs of specific customers. Product development managers in 12 companies surveyed about the reason for product development delays most often cited "poor definition of product requirements."[16] A sales force that will do anything to please a customer places a heavy burden on engineering and manufacturing in the redesign of existing products. This process problem can be addressed in several ways: by compensating the sales force on profit rather than revenue (because reengineering products is often not very profitable); by providing the sales people with automated tools that help them understand the implications of specification changes for product costs and delivery dates (this is being done by Bethlehem and several other firms in the steel industry[17]); by pre-engineering a set of component designs that can be combined into many different versions of a product (an approach that has been adopted by at least one firm in the group insurance industry and by two manufacturers of large trucks, Freightliner and Paccar[18]). The latter approach constitutes a cross-functional process in which neither sales, which tries to please the customer, nor manufacturing, which strives for efficient production, is wrong. A good product delivery process must combine both objectives and ensure that both functions understand the consequences of leaning in one direction or the other.

A number of firms attending to changes at the engineering/manufacturing interface have focused on reducing the number and impact of engineering change orders. In General Dynamics' Electric Boat Division, the change order process, which involves the customer (in this case, the U.S. government) as well as engineering and manufacturing, was one of the first processes to be redesigned. The division's great strides in reducing both the number of, and the manufacturing delays that result from, change orders were made entirely through organizational changes, without any use of IT.

Quality function deployment (QFD), a rigorous if somewhat unimaginative approach that has come out of the quality movement, can be applied to issues at the interface between engineering, manufacturing, and marketing. QFD relies on a series of matrices that relate desired product characteristics to design criteria and manufacturing requirements. Given that its greatest benefit is clarity in product specifications, which can eliminate much wasted time and design rework, QFD has been reported to be most appropriate in contexts where attributes involve incremental change and can be easily quantified.[19] Like most quality-based methods, it is less well suited to product development environments in which radical change or innovation are desired.

The many possibilities for innovation in design processes notwithstanding, Dixon and Duffey argue that there has been much more talk than action. They say that the U.S. "design infrastructure" of education, research, and company management is in disrepair and point out that most process interface improvements between design and manufacturing have been manufacturing initiatives.[20]

Yet many firms are reducing the time and the cost of product design and engineering by applying IT and organizational change enablers to the redesign of processes, targeting both those processes that are internal to engineering and those that cross functional boundaries with manufacturing, marketing, or sales. Their efforts set a good example for firms that are attempting to apply process innovation in other parts of the business.

MANUFACTURING PROCESSES

One might imagine that the blue-collar, shop floor world of manufacturing would be the last place in which process redesign would occur. In fact, we find just the opposite; manufacturing processes are the most likely source of innovation and process excellence in many firms, and have been the target of both technological and organizational change enablers. Core manufacturing processes are, on average, probably 10 years ahead of service or customer-facing processes in terms of the application of innovative process thinking.

The world of manufacturing process innovation is so advanced and wide-ranging as to be worthy of a separate book, but because the linkages between manufacturing and other functional processes are so important, we could not leave it out of this one. Those who would delve deeper than the overview given here might consult some of the other works cited in this chapter.

Although speed of production is to some degree an issue in manufacturing processes, several other process objectives are also of importance, chief among them the quality of the manufactured product or process output, which may involve a variety of product characteristics other than elimination of defects.[21] Others include: the flexibility to manufacture different types of products; reliable, predictable adherence to manufacturing timetables; and the lowering of the cost and price of products. All these objectives can be targets of process innovation and of specific change enablers (see Figure 11-4).

A recent survey of North American manufacturers suggests that time-oriented objectives, service, and flexibility are becoming more important in manufacturing processes.[22] Quality conformance, price, and overall product performance are less important now than they were in 1984, perhaps because these objectives are today taken for granted by customers.

Figure 11-4   Enablers of Innovation in Manufacturing Processes

- Linkages to sales systems for build-to-order
- Real-time systems for custom configuration and delivery commitment
- Materials and inventory management systems
- Robotics and cell controllers
- Diagnostic systems for maintenance
- Quality and performance information
- Work teams

In manufacturing as in other functions, the best way to achieve time, service, and flexibility objectives is through greater functional integration. As one book on world-class manufacturing puts it, "Matching marketing, engineering, and manufacturing capabilities is what differentiates world-class competitors."[23] We have discussed the interface between manufacturing and engineering in terms of design for manufacturability and reducing engineering changes in the previous section and will treat the manufacturing interface with logistical functions in the following section. Here, we shall deal with the process interface between manufacturing and sales.

To better coordinate manufacturing with sales, many companies are attempting to manufacture products and quantities to the specifications of customers and to minimize delay in delivery. This coordination, most pervasive in Japanese firms that serve domestic markets, is referred to in automobile manufacturing as the "lean production system."[24] Firms that employ this system manufacture cars to detailed customer specifications and deliver them in less than one week. To perform this feat, these companies must tightly integrate sales, manufacturing, and logistics into one smoothly functioning process. Although IT is applied at various points, technology is subordinate to organizational integration as a change enabler.

In consumer foods, the link between sales and manufacturing takes the form of deliveries to retail stores based on consumer demand. Frito-Lay has attempted to tie the quantity of snack foods manufactured more closely to the movement of the foods in grocery stores by using information from its sales force's handheld computers to drive manufacturing.[25] The firm has already used this store-level data to reduce by half the level of stale goods removed from stores.

Pepperidge Farm has linked its sales and manufacturing processes in an effort to meet specific freshness targets for all merchandise in all stores. The company has placed computers in the hands of its sales force, opened several new regional factories, and is pursuing tighter integration among the functions involved in the store replenishment process.

For other types of manufactured goods, manufacturing-centered processes increasingly involve providing a higher level of service to the customer, a trend Chase and Garvin have termed "the service factory."[26] Types of services targeted by this concept include delivery commitments, pricing, financing, and consulting. A service factory strives to be easier to do business with than its competitors, for example, by making and honoring real-time delivery commitments to customers, providing them information about how best to purchase and use products, and arranging optimum financing—usually with the aid of information technology. Among the firms explicitly pursuing this concept are Harley-Davidson and Digital Equipment Corporation.

Of the many manufacturing processes that have already been redesigned, those most likely to have been substantially improved are those that involve highly routinized, structured work flows. But many other processes have yet to benefit from redesign or the adoption of a process. The assumption in so-called job shop manufacturing, for example, has been that an emphasis on low volume, custom-built products precludes a focus on process improvement and innovation.

Yet job shops are like any other relatively unstructured process in which an outcome-focused view of work flow can yield significant benefits. A job shop division of Schlumberger, for example, had a factory that built 200 different oil field products in monthly volumes ranging from 1 to 20.[27] More than two years of radical process change produced remarkable benefits: a tenfold reduction in defect rates, a halving of the overhead to direct labor ratio, and 60% reductions in inventory. The division achieved these results not with IT (indeed, it backed away from several IT investments), but by switching from a batching approach characterized by lengthy lead times and queues to a cell-based work flow. Information—about product quality, adherence to schedules, and overall process performance—was a key enabler of the change.

Equipment maintenance is another relatively unstructured manufacturing process that might benefit greatly from an application of the principles of process innovation. Manufacturing personnel, particularly in high-technology industries, spend a great deal of time and money on maintenance. Innovations that reduce the time required to diagnose a complex equipment malfunction and recommend corrective action can speed not only the maintenance process, but other processes that depend on it. Digital Equipment Corporation, Schlumberger, Du Pont, FMC, and many other companies are pursuing this objective through the development of diagnostic expert systems.

Much of the potential for innovation in manufacturing processes has yet to be realized. Such important sources of innovation as robotics and cell controllers, materials requirements planning (MRP) and manufacturing resource planning (MRP II) software, and the desire to give workers more meaningful, varied jobs have all fallen into varying degrees of disrepute, and companies have retreated from using them as change enablers. In most cases, the fault has lain in too great an emphasis on technology and insufficient balance and process orientation.

Installation of so-called factory automation equipment such as robotics and manufacturing cell controllers peaked in the mid-1980s.[28] Many firms that viewed these technologies as panaceas failed to change overall work procedures or human resource practices. MIT researchers subsequently found that in the automobile industry, both in the United States and Japan, plants that added high levels of technology and did not innovate with human resources did worse than those that innovated only through human resource practices.[29]

Information technology for scheduling and resource management has long been viewed as an enabler of innovation in manufacturing processes. Indeed, the difficulty of managing plants with complex product lines and timely delivery requirements virtually necessitates MRP (for production control) or MRP II (for integrated production control and materials management). Yet many of the firms that have attempted to innovate with these systems over the past 20 or 30 years have failed.[30] We believe they failed because they viewed these technologies as solutions in themselves rather than as enablers of radical change. These firms did not address the entire process affected by the systems, and neglected to change associated subprocesses (inventory management, routing procedures, and so forth). Some simply lacked the discipline and sophistication to make use of such rigorous and complex tools. Their failure provides a painful but instructive example of the difficulties of IT-enabled process innovation, which offers lessons for other, nonmanufacturing processes.

IT is not the only enabler of innovation in manufacturing processes with which firms have had difficulty. One frequently discussed organizational innovation involves structuring work so that teams build entire products rather than relatively simple components. Although this concept, most commonly employed in Swedish automobile factories, has received much attention in the press, even as early as 1980 serious researchers questioned whether it was a clear answer to problems of manufacturing work design.[31] More recently, although Volvo and Saab continue to experiment with these approaches on a broad level, resulting productivity levels have met neither their own expectations nor the level demanded by foreign competition.[32] It may not always be possible to create more meaningful jobs for workers and achieve high productivity simultaneously. Ironically, at a time when this approach is being abandoned in manufacturing industries and functions, it is being adopted in the white-collar and service sector (see the discussion of case management in a later chapter).

Other technological enablers such as lasers, composites and ceramic materials, and precision machining equipment can open up new possibilities for process flexibility and performance. But issues related to organizational change and adaptation apply to these sophisticated manufacturing technologies, just as they do to information technologies.

Despite problems with key enablers of change, many companies have made substantial and even radical improvements in manufacturing processes. More than with any other large process, companies have tended to adopt a process view of manufacturing and to experiment with IT and other change enablers. Most of the credit for this sophistication attaches to the quality movement; quality precepts have been applied most frequently to manufacturing (see Appendix B). The challenge for companies striving to bring quality to manufacturing will be to integrate the notions of radical (as opposed to incremental) process change and to address not only manufacturing processes, but the interfaces between manufacturing processes and other key processes involved in product delivery. One of these is logistical processes.

LOGISTICAL PROCESSES

Logistical processes, those involving the movement of goods through an organization, are extremely important to successful product delivery. The past decade has seen considerable growth in the perceived importance of integrated logistical processes. Notions such as the "supply chain" are really just other ways of expressing the importance of a process view of logistics. Many process innovations in the manufacturing domain—just-in-time delivery of manufacturing supplies and components, for example—are logistics-oriented.

Success in logistics, according to one overview of logistical activities in manufacturing companies, largely turns on the application of a process perspective:

> A common thread runs through these situations where decisions along the logistics chain have resulted in a decided competitive advantage. These companies considered the management of materials in an integrated fashion, rather than from a functional "silo" perspective. Decisions were made, based not on functional performance or the cost/profit center concept, but on the logistics perspective of an integrated approach to satisfying the customer.[33]

Companies' efforts to innovate logistical processes—in sourcing, inbound logistics or supply chain management, configuration and outbound logistics, and third-party partnerships for inventory management and transportation—involve major uses of information technology and organizational change enablers (see Figure 11-5).

Many companies attempting to better coordinate the procurement and delivery of goods from suppliers along the entire supply chain are turning to JIT practices, a subset of supply chain management. A survey of 21 suppliers to Japanese-managed U.S. companies revealed that successful implementers of JIT achieved radical improvements in lead time, defect rates, and inventory

Figure 11-5   Enablers of Innovation in Logistical Processes

- Electronic data interchange and payment systems
- Configuration systems
- Third-party shipment and location tracking systems
- Close partnerships with customers and suppliers
- Rich and accurate information exchange with suppliers and customers

reduction.[34] Information systems that communicate and coordinate shipments and inventory levels are key enablers of JIT, and of supply chain management more broadly. Though it is possible to accomplish JIT without electronic data interchange, it is certainly much more difficult.

Effective management of supply chain processes often involves more than electronic communications. Becton-Dickinson's creation of a world-class supply-chain management capability, for example, required simultaneous changes in structure, skills, management systems, and other aspects of the organization.[35]

Routine communications both internal and external to the firm must also be managed. Helper points out that a rich flow of communications has been the most important contributor to better supplier relationships in the automobile industry.[36] In another industry, Hewlett-Packard, when it set out to redesign and radically improve relationships with suppliers, found the most common causes of problems to involve such mundane issues as misunderstood commitment dates, varying routing guides, and data entry errors.[37]

Communications requirements are greatly eased for firms that have fewer suppliers with which details and problems must be worked out. Supply chain processes are also easier to manage if a firm has considerable bargaining power relative to its suppliers, enabling it to force such supply facilitators as rapid, short-distance deliveries. Such conditions can be difficult to bring about; some defense industry firms, for example, are forced by their government customers to employ large numbers of suppliers. Companies that do enjoy leverage over suppliers should take advantage of it in the design of their inbound logistics processes.

On the outbound logistics side, firms are increasingly conscious of the long time lags that can be incurred and many errors that can be introduced in converting orders to delivered products. We have observed that, until they adopt a process view, many firms are unaware of the length of their order-to-delivery cycles. Rank Xerox, for example, finding that the process sometimes took as long as 40 days, introduced process innovations that have reduced it to 4 days for many products. Similar reductions have been pursued by Xerox in the United States.

Companies seeking radical improvements in outbound logistics must address a number of different aspects of the process:

- *Configuration.* Complex products with many versions and components are often difficult to configure. The world standard in the use of IT for configuration is Digital Equipment Corporation's widely publicized XCON system, which is still yielding benefits for its creators in terms of time, cost, and error reduction.[38]

- *Information management.* Information about delivery commitments is often subject to misunderstanding or miscommunication. One firm we studied used six different commitment dates in the outbound logistics process, none of which was the delivery date desired by the customer! Companies must be clear and firm about which dates really count, and eliminate all sources of information confusion.

- *Warehousing and finished goods inventory management.* The firms most successful at outbound logistics have totally eliminated warehousing of finished goods. They create finished goods only to fill customer orders and ship completed goods directly to customers. This involves new process behaviors on the part of both manufacturer and customer, who may not be used to drop shipping or receiving drop shipped goods.

- *Ancillary processes.* Many companies delay ancillary but important processes (e.g., credit checks, site preparation, and so forth) until the time of shipping. These activities should not be performed serially in the "critical path" time frame for moving goods to customers, but rather as parallel subprocesses.

Close relationships with third parties are central to the success of both inbound and outbound process innovation. On the inbound side, third parties are typically suppliers. In the retail industry, for example, leading firms are establishing tight partnerships with suppliers that involve supplier responsibility for managing all aspects of the supply chain, from reordering to shelf management. Wal-Mart and Procter & Gamble, in particular, are aggressively exploring concepts of continuous replenishment.[39] Other retail store chains, such as J.C. Penney and Federated Stores, are experimenting with turning entire store departments over to manufacturers that may install their own employees, inventory, and even point-of-sale systems.

In the manufacturing industry, suppliers can add value to logistical processes through such innovations as consolidation of purchased components from other third parties, kitting and packaging to suit customer requirements, and just-in-time delivery. Suppliers can also work with their manufacturing customers to eliminate the need for incoming inspection and testing of components. Finally, suppliers can provide information to manufacturers about the most effective use of their products, e.g., safety or quality data. This concept led Du Pont to be an early user of electronically provided Material Safety Data Sheets, which are furnished along with the chemicals it sells to its customers to ensure their safe use.

Third-party relationships on the outbound logistics side are most frequently with transportation and warehousing companies. Among the firms that have decided to outsource outbound logistics by forming partnerships with third parties are IBM, which has engaged Federal Express to manage and deliver spare parts for computers, and Ford, which has partnered with Ryder for distribution of completed vehicles.

Information technology advances made by third-party firms are partially responsible for highly efficient processes for outbound logistics. Federal Express (package express), American President Lines (ocean shipping), Schneider National (trucking), and CSX (rail) employ bar coding and scanners, satellite-based locational capabilities, customer premise terminals, and routing algorithms to deliver products reliably and predictably, qualities essential to customers that rely on JIT deliveries.

As transportation companies become more skilled in third-party inventory management (Federal Express, for example, has formed a division called Business Logistics Services), and as manufacturers decide to focus on core processes rather than those at the periphery of their businesses, business partnerships will become increasingly common.

SUMMARY

Although many examples of IT and organizationally enabled process innovation in product development and delivery can be found, there are many more companies in which just the adoption of a process view would be a major step forward. The list of companies making process innovations is often the same as the list of those making product innovations—Ford, Xerox, IBM, Hewlett-Packard, and a few pharmaceutical and defense firms.

It may be possible for smaller, less investment-oriented firms to adopt process innovations more easily than product innovations. Business processes can usually be identified and implemented at relatively low cost, and organizational change is easier the smaller the scale of the process and the organization implementing it. We noted several examples of environments in which heavy investments in technology were not necessary—or were counterproductive—to process success. Process innovation may be even more important than product innovation to the success of Western firms, and smaller firms may find it easier than larger ones to implement the organizational changes necessitated by process innovation. But no matter how small or unsophisticated a firm, product development and delivery processes can yield major improvements in business performance.

Although we have addressed a number of aspects of customer involvement in product development and delivery processes, the processes discussed here are generally not customer-facing. We now turn to marketing, sales, and customer service processes that are becoming a key focus for many firms.

Notes

[1] Edward B. Roberts, "What We've Learned: Managing Invention and Innovation," *Research Technology Management* 31:1 (January–February 1988): 11–29.

[2] Reported in Donald G. Reinerstein, "Who Dunit: The Search for the New-Product Killers," *Electronic Business* 9:8 (July 1983): 62–64.

[3] For example, a recent work on speeding product development makes no mention of research, although it is an excellent source in other respects.

[4] For a discussion of gate reviews, see Stephen R. Rosenthal, *Effective Product Design and Development* (Homewood, Ill.: Business One Irwin): 32–36.

[5]Thomas H. Davenport and Thomas Nickles, "The Strategic Use of Information Technology in Agricultural Chemicals," *Indications* (Cambridge, Mass.: Index Group, 1988).

[6]Philip A. Roussel, Kamal N. Saad, and Tamara J. Erickson, *Third Generation R&D: Managing the Link to Corporate Strategy* (Boston: Harvard Business School Press, 1991).

[7]James L. Nevins and Daniel E. Whitney, eds., *Concurrent Design of Products and Processes* (New York: McGraw-Hill, 1989).

[8]The Black & Decker example is briefly described in George Stalk, Jr., and Thomas M. Hout, *Competing Against Time* (New York: Free Press, 1990): 144–145, 148.

[9]For a discussion of the problems in utilizing CAD systems, see Jeffrey K. Liker, Mitchell Fleischer, and David Arnsdorf, "Fulfilling the Promises of CAD," *Sloan Management Review* 33:3 (Spring 1992): 74–86.

[10]Donna Stoddard, "Information Technology and Design/Manufacturing Integration," Ph.D. diss., Harvard Business School, 1991.

[11]Clarence L. "Kelly" Johnson with Maggie Smith, *Kelly: More Than My Share of It All* (Washington, D.C.: Smithsonian Institution Press, 1987): 97.

[12]G. Steven Burrill and Stephen E. Almassy, "Electronics 91: Framework for the Future," Ernst & Young, 1991.

[13]Ralph E. Gomory and R. W. Schmitt, "Science and Product," *Science* (May 27, 1988): 1132.

[14]See James W. Dean and Gerald I. Susman, "Organizing for Manufacturable Design," *Harvard Business Review* (January–February 1989): 28–36.

[15]Personal interviews with Xerox executives, June 1991.

[16]Ashok K. Gupta and David L. Wilemon, "Accelerating the Development of Technology-Based New Products," *California Management Review* 32:2 (Winter 1990): 24–44.

[17]Personal interviews with Bethlehem Steel executives in April 1991.

[18]The product development processes of Freightliner and Paccar are described in Stalk and Hout, *Competing Against Time*, 174–177.

[19]Smith and Reinertsen, *Developing Products in Half the Time*, 96.

[20]John R. Dixon and Michael R. Duffey, "The Neglect of Engineering Design," *California Management Review* 32:2 (Winter 1990): 9–23.

[21]See, for example, David Garvin, *Managing Quality* (New York: Free Press, 1988): 49–68, for a discussion of the multiple dimensions of product quality.

[22]J. G. Miller and A. V. Roth, "Manufacturing Strategies: Executive Summary of the 1988 North American Manufacturing Futures Survey," Boston University Manufacturing Roundtable, Boston, 1988.

[23]National Center for Manufacturing Services, *Competing in World-Class Manufacturing: America's Twenty-First Century Challenge* (Homewood, Ill.: Richard D. Irwin, 1990).

[24]The lean production system is most completely described in James P. Womack, Daniel T. Jones, and Daniel Roos, *The Machine That Changed the World* (New York: Rawson Associates, 1990).

[25]Nicole Wishart and Lynda Applegate, "Frito-Lay, Inc.: HHC Project Follow-Up," 9-190-191. Boston: Harvard Business School, 1990.

[26]R. B. Chase and David Garvin, "The Service Factory," *Harvard Business Review* (July–August 1989): 61.

242   *Product and Service Development and Delivery Processes*

27The Schlumberger case, and the overall call for innovations in job-shop manufacturing processes, are described in James E. Ashton and Frank X. Cook, Jr., "Time to Reform Job Shop Manufacturing," *Harvard Business Review* (March–April 1989): 106–111.

28These conclusions are supported by a 1988 capital spending survey of North American manufacturers by *Production Magazine*, reported in National Center for Manufacturing Services, *Competing in World-Class Manufacturing*, 306.

29Womack, Jones, and Roos, *The Machine That Changed the World.*

30National Center for Manufacturing Services, *Competing in World-Class Manufacturing*, 229.

31See, for example, the review of this topic in J. Richard Hackman and Greg R. Oldham, *Work Redesign* (Reading, Mass.: Addison-Wesley, 1980).

32Steven Prokesch, "Edges Fray on Volvo's Brave New Humanistic World," *New York Times* (July 7, 1991): F5.

33Christopher Gopal and Gerry Cahill, *Logistics in Manufacturing* (Homewood, Ill.: Business One Irwin, 1992): 13.

34Frank Barton, Surenda P. Agranal, and L. Mason Rockwell, "Meeting the Challenge of Japanese Management Concepts," *Management Accounting* 70 (September 1988): 49–53.

35Alfred J. Battaglia and Gene R. Tyndall, "Working on the Supply Chain," *Chief Executive* 66 (April 1991): 42–45.

36Susan Helper, "How Much Has Really Changed Between U.S. Automakers and Their Suppliers?," *Sloan Management Review* (Summer 1991): 15–28.

37D. Burt, "Managing Suppliers Up to Speed," *Harvard Business Review* (July–August 1989): 133.

38Dorothy Leonard-Barton and John J. Sviokla, "Putting Expert Systems to Work," *Harvard Business Review* (March–April 1988): 91–98.

39For a discussion from Procter & Gamble's perspective, see W. Weart, "Procter & Gamble's War on Paper," *Distribution* (May 1988): 88–89.

Chapter 12

Customer-Facing Processes

Processes that involve direct contact with customers traditionally have fallen into the functional areas of marketing, sales, and
service. Today, these processes cut across many functional areas.
Although any division of processes into a few categories is arbitrary, and even contrary to the benefits of process thinking, for
the sake of expediency we limit discussion here to processes that
involve direct contact with customers. But because even processes
that do face the customer depend heavily on other back-room
processes, including manufacturing, logistical, and financial processes, the customer perspective, either internal or external, should
pervade all processes. From the standpoint of the customer, a
process is only as good as its weakest link.

Although a process view of a business normally replaces functional perspectives as a means of organizing work, the most understandable way to describe processes is often in terms of the
functions they replace. Thus, in this chapter we discuss three
types of processes: marketing processes, designed to get customers
to initiate a relationship or transaction with an organization;
sales processes, the activities associated with customer purchase
and receipt of, and payment for, products or services; and service
processes that provide for post-sales maintenance of customer
relationships. In some organizations, these activities are best considered as one interlinked process; in others they can reasonably
be broken down into three or more related processes. But even
when viewed separately, they should still be linked by information flow.

Customer-facing processes that extend into the customer organization, such as order management and customer service, demand that interorganizational processes be treated as seamlessly
as possible and be designed jointly by supplier and customer. This
is, of course, quite difficult. Most suppliers have many customers,

and designing a single process that meets the needs of every customer is virtually impossible. Moreover, senior executives in a supplier firm may exert some influence over process behavior in their own organizations, but they have little influence over customer employees. Nevertheless, despite these constraints, joint investigation and decision making by supplier and customer should be a major objective in many of the process activities described below.

MARKETING PROCESSES

Marketing-oriented processes increase the likelihood that a customer will engage in a transaction or relationship. But the open-ended nature of marketing activities makes it difficult to know, at least immediately, whether a particular set of activities results in an actual transaction or relationship. The primary output of a marketing process is thus highly uncertain, and this uncertainty probably accounts in large part for many companies' unwillingness to consider marketing activities in process terms. That marketing people frequently consider themselves to be creative and their work unstructured further discourages the adoption of a process orientation.

Some of the activities that comprise marketing may be more amenable to a process view than the marketing process overall. A process orientation can benefit such activities as market selection and definition, customer and market information collection and use, and marketing-initiative planning and tracking. Clearly defining the inputs to and outputs of these activities, describing their flow, and measuring their performance brings rigor to processes that are sometimes poorly defined. Figure 12-1 lists specific tools and enablers that can be applied to marketing processes.

A self-described marketing-oriented firm that employs many people in marketing activities, IBM has aggressively targeted marketing processes for redesign. Three of eighteen key processes being redesigned by the company are marketing processes. Key objectives of these innovation efforts are better definition of market segments and the products and services directed at them and expanded capture and use of information about customers.

But the industry that has made marketing virtually a closed loop process—one in which the results of marketing initiatives

Figure 12-1   Enablers of Innovation in Marketing Processes

- Customer relationship databases/frequent buyer programs
- Point-of-sale systems tied to individual customer purchases
- Expert systems for data and trend analysis
- Statistical modeling of dynamic market environments
- Close linkages to external marketing firms

can be measured and understood—is consumer goods. A combination of intense marketing pressures, often manifested in product promotions and heavy advertising, and recent innovations in point-of-sale information technologies, such as scanners, has made it increasingly possible to assess the impact of marketing initiatives accurately. Test market use of "people meters" to compare what the advertising consumers see with data about what they buy in supermarkets can help companies determine how well their advertising is working and provide almost immediate feedback on how a promotion is affecting sales. Several firms, including Philip Morris and Quaker Oats, have national databases of their customers and are beginning to understand the patterns of customer purchase behavior for their products.[1]

As firms begin to compile data on their customers, they can start to manipulate marketing processes at the individual or household level. But these processes are quite complex, involving analysis of individual purchase history, store-level analysis, pricing and promotions, product variations, and alternative marketing channels. Understanding and manipulating these variables requires development of complex statistical models. If these models are built and used, marketing can become more of a science than it is today.

However, given the nature of the people attracted to marketing and advertising, not even the most sophisticated marketers in consumer goods firms have explicitly recognized the potential for applying process structures and discipline to their work. Were they to do so, their technological innovations might yield more effective processes. In most firms, problems of too much data, managers who do not understand the information or the technology, and fear of radical change have limited the effects of a closed-loop process. It is clear, however, that over the long run the consumer goods industry will revise its approaches to marketing and selling, and other industries may follow its example.

Particularly in the consumer goods industry, the sheer volume of marketing data constitutes a barrier to its effective use. One provider, Information Resources, collects about two billion characters of data per week. Even the most sophisticated firms can use only a small fraction of available data. To facilitate analysis, both provider (e.g., Information Resources) and user (e.g., Frito-Lay) firms are developing systems that employ algorithms or expert logic to identify exceptions to normal patterns in data. One system, used by several firms, including Ocean Spray Cranberries, even generates textual and graphic reports from marketing data without human intervention.[2]

IT is improving understanding of buyer behavior in other industries as well. Sophisticated telecommunications firms and airlines are beginning to take advantage of customer transaction data for marketing purposes, for example, using frequent buyer programs to associate buyer behavior with individual customers so as to be able to target marketing initiatives at the best prospects. Such "narrowcasting" employs direct mail, individualized magazine publishing, telemarketing, and better informed use of television advertising time.

Because much of the relevant expertise and information resides with third-party firms, companies that want to improve their marketing processes must develop close partnerships with advertising agencies and with data collection and database marketing firms. Such partnerships promote faster flows of more useful marketing data to decision makers through joint participation in such activities as design of data collection methods, development of analysis tools, and even marketing of new tools and techniques. Process performance improvements are not likely to accrue to a company content to be just another transaction-oriented customer of these third-party firms.

Order Management and Other Sales Processes

Many process-oriented firms have focused their redesign efforts on order management, an extremely important process that encompasses a range of activities from preparing sales proposals to billing and collection for goods and services provided. Although we focus here on order management, other sales processes warrant consideration for redesign and innovation, including sales coverage planning, customer account management and planning, lead and prospect management, and management and tracking of the sales force and sales results. Key enablers currently being used to innovate sales and order management are listed in Figure 12-2.

Order management, or order fulfillment, is the heart of sales processes. It extends from the time a customer places an order with a firm through receipt of the order by the customer to receipt of payment by the seller. Although it may take only a short time, this process involves the coordination of such diverse activities as credit checking, manufacturing, logistics, accounts receivable, and even relationships with external suppliers for build-to-order

Figure 12-2   Enablers of Innovation in Sales and Order Management Processes

- Prospect tracking and management systems
- Portable sales force automation systems
- Portable networking for field and customer site communications
- Customer site workstations for order entry and status checking
- "Choosing machines" that match products and services to customer needs
- Electronic data interchange between firms
- Expert systems for configuration, shipping, and pricing
- Predictive modeling for continuous product replenishment
- Composite systems that bring cross-functional information to desktops
- Customer, product, and production databases
- Integration of voice and data (e.g., automated number identification)
- Third-party communications and videotext
- Case management roles or teams
- Empowerment of frontline workers

products or shipping, which normally take place in several different business functions (see Figure 12-3).

Order management, being a primary determinant of customer satisfaction, is of critical importance to many firms. If customer collections are included in the process (and they should be), it is the key factor in cash flow. But the many linkages between functions involved in order management allow multiple opportunities for work to fall between the cracks, and therefore make the process a strong candidate for redesign.

Redesign of order management processes is being driven heavily by customers that want speed, reliability of commitment, instantaneous access to order status, and no mistakes, and that want to receive products or services using innovative and efficient processes of their own. Moreover, sellers increasingly are being called upon to deliver products and services without adding inventory or excess capacity. Wal-Mart has even demanded that its vendors do not employ middlemen or distributors. These objectives may seem daunting, but many companies are beginning to meet them.

Companies are adopting several different strategies for effecting major process innovation in order management. We describe these below.

Case management. Firms in a number of industries that formerly had complex, multifunctional processes for order management are consolidating the order management process in one individual or team, wherein the case manager works in a closed loop process that handles an entire customer transaction from beginning to end. Case managers typically have access to cross-functional information systems with data and applications. We describe this notion in greater detail below.

Figure 12-3 The Order Management Process

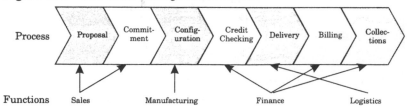

Order segmentation.   Many firms are finding that they can create more efficient processes if they categorize orders by complexity or difficulty. Straightforward orders can often be handled by computer. IBM Credit Corporation, for example, currently handles 55% of all requests for custom financing bids by computer and completes them almost instantaneously. The remainder are processed by case managers.

Customer participation.   There are many precedents for letting customers manage parts of the order management process. The most common cases involve customers entering and tracking their own orders. Examples include Baxter Healthcare's ASAP system[3] and Federal Express's Powership customer-premise terminal. More sophisticated versions of customer participation take the form of customer configuration and scheduling of orders and shipment. American President Lines is attempting to become, in its own words, "customer managed." Customers can reserve space on APL ships, calculate shipping costs, and check shipment status and arrival dates.[4] A related version of customer process management that pushes as much of the process as possible out to sales representatives may be a transitional state on the way to customer management. The difficulty with this orientation often lies in persuading customers to adopt process innovations in which they must participate. Baxter, for example, has developed sophisticated order management capabilities through six generations of the ASAP system, yet many of its hospital customers continue to use its second- and third-generation systems.

Real-time process execution.   At the highest level of customer service, order management demands real-time performance. Price and ship date commitments, for example, should be made when the customer places an order. This is usually possible only with sales force computers that support access to inventory databases and pricing algorithms. A major food distribution company, for example, used laptop computers to enable sales representatives to confirm food availability and price for institutional buyers and guarantee next-day delivery, thus allowing customers to plan menus without fear of missing ingredients.[5] Although the process worked well in a pilot implementation, some representatives found the technology intrusive when used in the presence of customers—the software didn't work as expected, or a communications connection couldn't be established. As a result, the system was not implemented broadly. The same company is now investigating customer-premise systems that would eliminate order taking by sales reps. In the consumer financial services industries, the standard for customer-facing process performance is increasingly real-time execution of the entire transaction at the customer site. This is already a goal of a number of insurance, leasing, and mortgage companies.

Parallel processes.   As we have noted, it is often possible to introduce some degree of parallelism in serial process. The most common example of parallel task performance in order management is separating credit checking and other financial processing from the rest of the process. Parallelism is a particularly important tactic for order management when the goal is real-time execution.

Process partnerships.   Also discussed earlier was the growing number of firms forming partnerships to facilitate various aspects of order management by, for example, eliminating unnecessary or redundant transactions, exchanging work better suited to one partner than another, or changing restocking or payment triggers. This strategy allows firms to concentrate on the processes of critical importance to their success. Wal-Mart, a notable proponent of this process strategy, relies on its suppliers to continuously replenish, and manage shelf space for, their own products. We are also acquainted with a leasing firm and insurance firm that form partnerships with their customers for mutual maintenance of billing information. Process partnerships can sometimes be beneficial to buyers, but expensive for sellers, and each partnership with a customer can be a source of nonstandard order management processes.

IT ENABLERS OF INNOVATION IN ORDER MANAGEMENT

There are at least two issues associated with IT-enabled process innovation in sales. One, many sales representatives, like most marketing personnel, view their work as being unstructured and resist process characterizations of it. Moreover, sales is often less welcoming than other functions of IT-based work changes, viewing laptop or handheld computers, for example, less as an aid to their jobs than as intrusive in the sales process, particularly when used in the presence of customers. Two, sales representatives may believe that the primary value of such technologies is to sales management rather than the sales force. When the primary sales application is, for example, call reports, such cynicism may be justified.

Databases. The sales force, being the primary source of customer information, can be a vehicle for improving certain marketing processes. In many companies, sales representatives "own" customer information, gathering and storing it as they see fit. Information technology, in the form of customer databases, can be used to steer information storage and retrieval in the opposite direction, making customer information an asset of the firm rather than of individuals alone. If such systems are to be successful, however, there must be benefits to the collector.

Electronic data interchange. Because order management is an interorganizational process, electronic data interchange is a primary enabler. Data can be exchanged electronically between firms and their customers and suppliers, and internally among design engineers, manufacturing, and logistics. Interorganizational data transfers are more constricting because they must adhere to standards of data content and format; internal communications can be more loosely structured. Little competitive advantage can be gained through the use of EDI, although, when combined with changes in buying and selling processes, the technology can yield substantial efficiencies and increases in speed. Westinghouse Electric, for example, had a team of consultants analyze the purchasing process of a major customer, Portland General Electric (PGE). The team considered both EDI and new purchasing processes as vehicles for improving PGE's purchasing efficiencies. It determined that EDI could save the customer approximately 10% over nonelectronic purchasing, but that a radical redesign of the purchasing process could save as much as 90% over existing processes. Because the PGE manager did not own the entire purchasing process, however, and because he did not feel empowered to change it, PGE decided to implement an EDI link with Westinghouse rather than attempt to implement major process changes.

Expert systems.   Expert systems, another key enabler of innovation in order management, can make internal expertise ubiquitous and instantly available. Expert systems are variously used to aid credit checking, pricing, and product configuration, and to resolve order management problems so complex that it is difficult for any but the most sophisticated employees to deal with them.

Digital Equipment Corporation has applied the expert systems technology throughout the order management process, from order entry to configuration. Some applications in this process have been unalloyed successes, for example, the XCON configuration system, which has saved Digital many millions of dollars and many expensive configuration mistakes.[6] Yet recent research at a firm resembling Digital suggests that not all aspects of the order management process have been successfully enabled by IT. Despite careful approaches to implementation, the more customer-facing uses of expert systems—for example, an order configuration system for the sales force—have not been successfully integrated into the company's sales force.[7] Digital, which has an expert sales configuration system called XSEL, has created a role separate from the sales representative's job for the configuration and use of XSEL.[8]

Field technologies.   Many firms have radically transformed processes for order management through the use of field technologies for data entry and access. These technologies include handheld and laptop computers, some with pen-based input devices that enable field personnel to enter and review information without typing. Many firms initially implement such technologies to promote faster, more accurate transfer of order information to manufacturing; it is only as they gain experience with these tools that they begin to consider their potential to effect changes in internal and interorganizational processes.

Frito-Lay's introduction of handheld computers to its route sales force provides an excellent illustration of this evolution. The company's original objective in equipping its 10,000-person sales force with handhelds was to accommodate increasing numbers of products by eliminating complex paper order-entry forms. As the data began to flow in from the field, sales and marketing managers began to use it to evaluate the results of promotions and new product introductions more rapidly. Even later, manufacturing began to use the information to reduce overproduction and the incidence of stale products. The company subsequently initiated a comprehensive effort, called "Project Pipeline," to redesign its entire order-management process.[9] Although current profitability pressures are slowing the move toward a fully integrated order-management process, the company has made great strides.

A computing platform in the field can also be used to radically expand the role of the sales representative. Colgate-Palmolive, for example, used laptop computers to change the role of sales representatives from order taking to profitability analysis and logistical planning. Sales reps can use the laptop software to calculate order profitability and shipping costs and arrange shipping, all of which have become their responsibility. Because they are compensated on the basis of order profitability as well as volume, it is in the sales reps' interest to renegotiate orders that are unprofitable because of uneconomical quantities or inefficient shipping methods.

We are only beginning to understand the impact of portable technologies on order management and the role of customer-oriented personnel. Most companies' field organization structures are predicated on the assumption that certain information-processing activities must occur in field offices. The existence of highly capable information technologies calls this assumption into question. Firms such as AT&T are beginning to close field sales offices and increase their reliance on laptop-equipped sales representatives who work out of customer sites and their homes. New user-interface technologies, including pen and voice input, and portable networking technologies employing packet radio and reliable cellular modems will eliminate the remaining technological impediments to transforming field-based order management.

In some industries, sales force field technologies are but a transitional step toward full networking with customers. Among the many examples of customer premise systems, some have achieved legendary status in the annals of strategic systems. Most, including the Baxter/American Hospital Supply, McKesson, and airline reservation systems, are proprietary and employ company-specific data structures and communications protocols. But the trend in customer-premise systems is to create industrywide systems, which can be shared by multiple companies that comply with common EDI standards. A number of the legendary company-specific systems, including Baxter and several of the airline systems, are moving in this direction.

Infrastructure systems.　Although little competitive advantage attends the use of industrywide systems, there are opportunities for one company to better integrate the order data or to use more of the information than another. Thus, Max Hopper, a key developer of the American Airlines' SABRE reservations system, although he has observed that there may be no further strategic systems in the airline industry, continues to pursue business process innovation.[10]

Many of the technologies that enable order-management process innovation help create not just a single application, but an entire infrastructure of technical capability. Such tools as databases, networking, composite information systems, and integrated application packages serve a variety of process needs. Databases of many kinds abound in successful order-management processes. They contain information about customers and their requirements, product characteristics, manufacturing schedules, and so forth. Coupled with expert system or algorithmic logic, they can be used to predict customer orders, immediately determine delivery dates for products, generate automated proposals, or target products or services to particular customer needs. Predictive modeling systems are commonly employed in continuous replenishment environments, such as those maintained between Wal-Mart and such suppliers as General Electric and Procter & Gamble.

To be adaptable to multiple process needs, a company's data structures must be simple and robust and organized by key aspects of its business. Frito-Lay's databases, for example, are structured by customer, product, and geography. The simplicity of this structure has enabled the company to make data readily accessible at multiple levels of the organization and use it across multiple functions. Because no manager could possibly analyze all the data from all locations every day, data analysis at Frito-Lay operates on the basis of exceptions to plans or patterns.

Networking and communications.　Communications networking is an extremely important—some have argued the most important—part of a company's process information infrastructure.[11] Once a network is established, a company often discovers many possible process innovations that might result from its use. Retailers such as K mart, Wal-Mart, and J. C. Penney, for example, have discovered high-capacity network access to remote store locations to be a prerequisite of effective order management and rapid response to merchandising trends, and consequently they have developed extensive networks using both satellite and ground communications technologies.

Another key aspect of networking capability, the ability to integrate voice and data in order management applications, is absolutely critical to fast customer response in telemarketing environments. The PC/Mac Connection, a fast-growing computer supply mail order firm in New Hampshire, uses automatic number identification (ANI) of incoming calls together with a customer database to achieve one-minute transactions for routine purchases. Many banks now offer detailed account information, and some account manipulation, through the telephone keypad.

A number of companies offer order management infrastructures that can be used by firms that lack the size or demand to warrant building their own infrastructures. The most prominent of these, and the one with the largest subscriber base at present, is Prodigy, a partnership of IBM and Sears that enables small companies to innovate in order management with little investment in technology. A flower delivery service using Prodigy, for example, can take advantage of the service's video imaging capability to picture alternative flower arrangements available to customers. Although not yet profitable, the service has grown much faster than telephone-based service because of its ability to picture products. The Minitel videotext system in France is more broadly institutionalized than any U.S. infrastructure capability.[12]

Cross-functional systems.   Successful use of IT for order management innovation usually involves substantial integration across functional systems. Companies that do not want to develop a new set of fully integrated applications have two options in this area: (1) to make use of composite systems that draw data from a variety of different systems and integrate it at the user level, and (2) to purchase an integrated applications package for order management.

Composite systems, also known as "frontware," are an increasingly common component of redesigned order management processes.[13] They cut across existing applications and data architectures, extracting and pulling together at the order management workstation specific data elements needed by the individual or team managing the order. A composite system for underwriting and issuing an insurance policy, for example, might draw from actuarial, medical records, credit, underwriting, and customer databases.

The use of composite systems, although a valid short-term solution to problems of integrating order management information, warrants caution over the long term. With more and more composite systems in a company's information architecture, the web of intersecting applications and data becomes highly complex. Changing a core system may necessitate changes in every composite system that draws data from it. Moreover, to the extent that new development efforts are expended primarily on composite systems, there may be insufficient resources to modernize and better integrate core systems. It is all too tempting for a firm to apply successive composite system band-aids and never fully convert to a process-oriented architecture.

Installation of an integrated applications package is a more long-term and investment-oriented approach, but only a few vendors offer such packages. Dow Chemical and Kodak are installing order management systems that combine the functions of order entry and processing, sales, distribution, invoicing, materials management, production planning, costing, and pricing. Although appealing, integrated order-management packages are not without problems. The systems are expensive compared to composite systems, and a company that purchases such a package must be willing either to adopt all of the processes assumed by it or expend substantial resources modifying it. Most companies have difficulty adopting a package that supports only one function; multifunctional packages are exponentially more difficult to assimilate.

The same cautionary messages that apply to IT enablement of innovation in other processes apply here as well. Companies should not assume that an IT-based solution, in and of itself, will change the order management process. The technology must fit the design of the process and be employed in conjunction with organizational and human resource and other types of enablers.

ORGANIZATIONAL/HUMAN RESOURCE ENABLERS OF INNOVATION IN ORDER MANAGEMENT

Because many of the IT enablers of innovation in order management involve changes in individual roles and authority, associated organizational and human resource initiatives must also be implemented. Here, we consider such key enablers as empowerment of frontline employees, the use of self-managing teams, and changes in organizational structure.

Empowerment. Well before the concept of IT-enabled process innovation was advanced, progressive management experts were urging empowerment of frontline employees who serve customers.[14] Frontline workers, they believed, should be able to take whatever steps necessary to fill orders in a way that satisfies customers. This, together with an orientation toward other process objectives such as time and cost, is the primary thrust in innovative order-management processes as well.

When order management is performed by a single individual or a team, empowerment must be cross-functional. That is, each function involved in a process has to rely on someone outside that function to execute the function's activities. Those who perform a process often do not "work for" the managers of the functions. Therefore, the culture must be one of facilitative management, in which trust in process performance is extended whether or not direct management control can be exercised.

There are also informational aspects to frontline empowerment. To provide the information employees need to serve customers well, an organization must place a high degree of trust in its employees. Customers, and information about them, are many businesses' most important asset. To entrust this information to a small group of employees relatively low in the hierarchy requires an extraordinary degree of empowerment and trust. Even when the "rules" for customer interaction are embedded in a system, the organization must trust employees to follow the system's recommendations and override them when necessary.

It cannot be expected that empowerment will simply emerge in an informal way; it must be explicitly planned and managed. When Xerox redesigned its order management process, for example, it created a list of specific "empowerments" for frontline personnel. These workers were generally granted all decision rights that did not have negative consequences in other regards. For example, pricing flexibility was very broad, but no price could be given that was below the price levels for government customers (a policy mandated by the government). After these empowerments were agreed upon, the firm then examined its previous policies and procedures, and pared away those that were incompatible with the empowerments list.

When frontline employees' jobs are changed, other aspects of the organization must be changed to accommodate the new roles that emerge. These changes, which include the development of new criteria for evaluation and compensation, can themselves be empowering. Because orders are often key to a firm's profitability, some firms have given sales or other frontline employees responsibility for ensuring profitability at the order level. When compensation is based on profit, the behavior of frontline personnel can change dramatically. To turn down a customer order because it is unprofitable may entail a radical change in corporate culture, but such a practice has been the norm in some industries (e.g., insurance) for many years.

Team-based structures.   Order management can be a fertile environment for enablement by team-based organizational structures. Order management is performed most effectively, because of the many functions involved, by cross-functional teams. Such teams provide social interaction for their members and the potential for a higher level of process expertise and execution. Image and call distribution software and groupware permit work to be distributed appropriately across members of a team.

Of course, there are teams and there are teams. Some of the "teams" we have observed in order management processes strain the definition of the term. Certain members of the team may drop in only once a day or week to solve problems requiring expert knowledge. One insurance firm designed an order management process in which members of the teams were scattered geographically and expected to communicate only by telephone or computer. This type of team provides only rhetorical benefit.

Perhaps the simplest organizational enabler of order management is assignment of ownership of the process to a key executive. Although not necessarily within the context of process innovation, many companies have begun to give order management a home within the organizational structure. Ownership is most frequently given to the customer service or integrated logistics functions.

Although it is too early to know the results of such structural change, we anticipate that, lacking a specific process context, it will be difficult to achieve consensus on ownership of the order management process. For example, making the customer service function more influential, as many firms have done, creates the possibility for "channel conflict" with the sales organization. Having the customer service function take on order processing may prevent a move toward a more automated version of order management. Empowering a single function may amount to no more than a band-aid, which provides only temporary and superficial benefit to the overall process.

BARRIERS TO INNOVATION IN ORDER MANAGEMENT

Among the barriers to innovation in order management that we have encountered most frequently is multiple financial and control systems. It is difficult to standardize, never mind innovate, an order management process when the financial component of order management varies widely by customer, product, division, and so forth.

One manufacturing company we studied had 37 different billing systems, according to the firm's head of customer administration, so that each organizational unit could better understand its contributions to the firm's revenues and profitability. No common system would satisfy these needs, so each unit created its own. The firm, not surprisingly, is having great difficulty even streamlining its order management process.

Companies are increasingly entering into complex, long-term relationships with varying legal arrangements. IBM provides a good example of this issue. Of three IBM sales representatives interviewed at random, one was trying to arrange a joint venture with local hospitals to offer imaging services for medical records, the second was attempting to establish a joint venture with a turkey processor in which IBM would supply robotic equipment for plucking and processing, and the third was proposing an outsourcing arrangement whereby IBM would hire many of the customer's information systems employees. Such arrangements may be good business deals; they may even be the price of doing business in the 1990s; but for the designers of the order management (called "customer fulfillment") process within IBM, these partnering relationships represent a process standardization nightmare.

In the right circumstances, supplier partnerships can be enablers of innovation in order management, but at other times they can actually constitute a barrier to it. In information technology and several other industries, firms are increasingly becoming "integrators" of other firms' products. This has long been the case in the automobile industry and other manufacturing environments in which products and components are standardized. But integrator businesses encourage customers to have it their way. With each product unique to a specific customer, businesses are faced with the need for process coordination and rapid information exchange among several companies. It is difficult enough for a company to innovate its own order management processes. When each order involves several different suppliers, the difficulty escalates dramatically. We know of no integrator businesses that have yet created efficient, effective order-management processes.

THE RISE OF THE CASE MANAGER

Many of the companies that have redesigned their order management processes to be fully executed by one individual or team have used the term "case manager" to describe this radical new role. This phenomenon is interesting because of its broad application and potential to become an entirely new organizational form. Case management can be applied to other processes (potential applications in service processes are described below), but the industries in which we have seen this type of design thus far have all applied it to the processing of customer orders of some type.

We have observed case management in a variety of white-collar and service industries—underwriting in insurance, circuit provisioning in telecommunications firms, patient management in health care, and customer relationship management in banking. Even manufacturing firms are experimenting with case management for filling orders. Companies that have decided to redesign their business processes at the customer interface are very likely to create a case management role.

Components of successful case management. Case managers are surprisingly alike across different companies and industries. We have observed four key components to successful case management:

- a closed loop work process that involves completion or management of an entire customer deliverable;
- role expansion and empowerment to make decisions and address customer issues;
- access to information throughout the organization; and
- a location in the organizational structure at the intersection of various functions or matrix dimensions.

In the absence of just one of these components, case management may not be effective. Several banks, for example, that constructed integrated customer files achieved little cross-selling or relationship management because they did not empower users to represent all bank products. Similarly, a high-tech manufacturer we studied created a case manager-like role to handle the order fulfillment process for key accounts, but because managers of relevant functional groups insisted that the role be nonexempt (and thus low in status), it was never given access to key information that it was intended to coordinate. Less than all might as well be nothing in case management.

The organization of work. However routine it might seem in many firms, case management is a radical departure from the way we have organized work since the Industrial Revolution. For Weber and Durkheim, classical organizational thinkers of the nineteenth century, the division of labor was a key aspect of modern organizational life. Adam Smith gave the following description of the division of labor with respect to the manufacture of pins.

> One man draws out the wire, another straights it, a third cuts it, a fourth points it, a fifth grinds it at the top for receiving the head: to make the head requires two or three distinct operations: to put it on is a particular business, to whiten the pins is another

> . . . and the important business of making a pin is, in this
> manner, divided into about eighteen distinct operations, which
> in some manufactories, are all performed by distinct hands,
> though in others the same man will sometime perform two or
> three of them.[15]

Experimentation with more complete jobs in manufacturing
began many years ago, but few firms have had overwhelmingly
positive results compared to the assembly line. For routine, structured tasks, it is difficult to beat highly specialized jobs and sequential organization for efficiency, and there is evidence that
workers (at least in the United States) who hold these jobs are
reasonably satisfied with them.[16]

Although many service industries adopted the assembly-line
model, it never worked as well in service as in manufacturing.
White-collar workers tended to be uncomfortable with the arrangement, and each highly segmented role created the need for
a buffer (inbox and outbox) and communications interface, which
increased greatly the time required to deliver the service. In many
insurance firms we have found 10-to-1 ratios between elapsed
time and value-added time in work processes.

Customers, who have a way of interfering with the best-laid
plans for how work should flow, are perhaps the biggest monkey
wrench in assembly-line processes. They call about their applications or policies or orders in midstream, forcing a halt in the
flow while their file is found. The products they requested don't
fit the firm's categories. They want their orders expedited, which
slows down other orders. For all these reasons, firms tried to
isolate frontline people from the back-office producers. For these
reasons, too, the influential organizational theorist J. D. Thompson counseled that those on the periphery of an organization, that
is, those involved with customers, should be "buffered" from those
at the core.[17]

But this front-room–back-room separation no longer works
in the business environment of the 1990s. Organizations that
want to be customer-driven can hardly consider customer interactions an interruption in their work. If every customer interaction is, as the SAS chief executive Jan Carlzon puts it, a "moment
of truth," then we must empower each customer-contact employee
to act for the entire firm. Information technology affords firms
the potential to view their entire organization from the front

line—why not realize that potential? Finally, competitive and financial pressures are driving many firms to pursue radical innovations in work processes in order to achieve radical performance improvements.

Case management makes the back room and front room indistinguishable. It is a vehicle for literally presenting "one face to the customer"—long a slogan in many firms, but rarely achieved when the "case" or customer deliverable is spread among many different departments and workers. Potentially gone are the delays and errors occasioned by front-room–back-room handoffs. The combination of a high level of consistency and efficiency and the ability to bring the expertise of an entire organization to bear through information systems is difficult to resist.

Beneficiaries of case management.   Employees as well as customers stand to benefit from case management. Case managers derive satisfaction from seeing a task through to completion and self-esteem from being trusted by their employer to make the right decisions and take the right actions. They are permitted to view a broad range of operational processes and face a variety of issues and challenges. Their jobs are not easy, but they are rarely uninteresting.

What kinds of organizations and processes can benefit from case management? We believe that organizations that rely on complex processes for creating their products and services and bringing them to market are the best candidates for case management. In manufacturing, case management probably works best when orders require complex interaction with customers—that is, when there are many products to choose from, when products must be configured to customer specifications, or when services are combined with products. Service industries become candidates for case management when the services they provide cannot be performed simply and straightforwardly, in assembly-line fashion, either because customers frequently interrupt the service operation to check its status or because multiple types of services are offered to individual customers, perhaps simultaneously (as in a health care institution). Such service industry environments might be termed "service job shops."

Implementing case management.   Case managers can be employed to overcome organizational complexity as well as complexity in products and services. When a customer must interact with multiple functions or divisions, a case manager can manage the interactions. In an ideal world, organizational handoffs in customer-facing business processes would be transparent to the customer; in this world, this is seldom the case.

Case management may not be able to deal with the highest levels of complexity; one insurance firm we studied had experimented with case management, but had later adopted a more segmented division of labor for its most complex processes. It seems well suited, however, to processes that involve moderate to high levels of product, service, or organizational complexity.

Companies attracted to case management must decide how far they will take the design. A key decision is how far into the field to extend case management. In the examples discussed thus far, the case management role has been occupied by a field salesperson, selling agent, broker, or other intermediary who deals directly with customers. Thus, case managers are typically one step removed from the customer, though they sometimes interact directly with the customer when they provide customer service.

Some firms have been bolder: Lithonia Lighting's Light*Link system places relevant information in the hands of independent sales agents who are empowered to act on Lithonia's behalf with customers. These agents, viewed by the company as being at the center of all information flows, have access to a great deal of information about Lithonia's internal operations, including production scheduling, pricing policies, and so forth. Moreover, Lithonia has placed at their disposal a number of expert systems that enable agents to respond quickly to customer requests, be they for an office building or a baseball park.

Similarly, although IBM Credit Corporation's creation of a case management role at headquarters greatly improved its quote generation process, the company believes that to compete successfully it must move quote generation out to the IBM field sales force. These account executives would seem to have every incentive to give away the store to land a customer's order, yet preliminary results find sales people arranging deals more favorable to IBM Credit than its headquarters staff was able to arrange. Some field sales representatives are even experimenting with turning over some aspects of quote generation—specifically the iterative "what if" analyses of financing alternatives—to their customers.

Less-bold firms, among them a number of insurance companies (with captive agents) and manufacturing firms, hope first to demonstrate the success of case management at headquarters. They may then gradually move the role and its capabilities out to the field.

Other key issues that confront firms trying to establish a case management environment for their customer-facing business processes include level of control, information architecture, and the choice between individuals and teams. Moreover, establishing a case management role and initiating case management activities has implications for human resource requirements and policies. There is too little experience with case management to know the right and wrong of dealing with these issues, but every case management process we have observed is working well. We view case management not only as a solution to order management process issues, but as a bellwether of a new organizational form.

SERVICE PROCESSES

Earlier, we considered the development and delivery of services, particularly with respect to logistical issues, but we have yet to address the broad range of post-sales service processes at the customer interface or the many service issues that face product companies. Products, for example, must be installed, maintained, and repaired. With the service that accompanies products becoming a major source of competitive differentiation, all industries are in a sense becoming service industries.

High service-business profit margins are forcing more and more product companies to take seriously the service end of their business. Such computer giants as IBM, Digital Equipment Corporation, and Unisys are examples. In fact, we have found product companies to be more process-oriented than service companies (with the possible exception of freight transportation and a few other service industries). This is perhaps because of the absence in service businesses of a manufacturing function, from which a process orientation might reasonably have originated as a by-product of the quality movement.

There is no shortage of process innovation objectives for firms in either type of industry. The usual goals of time and cost reduction are as valid for services as for products—often more so because they may be more visible to the customers. Nor is there a shortage of enablers of innovation for service processes (see Figure 12-4).

Figure 12-4   Enablers of Innovation in Service Processes

- Real-time, on-site service delivery through portable workstations
- Customer database-supported individual service approaches
- Service personnel location monitoring
- Portable communications devices and network-supported dispatching
- Built-in service diagnostics and repair notification
- Service diagnostics expert systems
- Composite systems-based service help desks

For evidence that service quality is becoming an increasingly important competitive differentiator, we can turn to the PIMS corporation database, which reveals that firms with high relative quality, whether in product or service industries, incur costs that are no higher, yet can charge approximately 5% more than their competitors.[18]

A long-term study by the Marketing Science Institute suggests that there are multiple dimensions to service quality, leading to multiple potential process objectives for service processes.[19] These dimensions include:

- tangibles (the appearance of personnel, facilities, and so forth);
- reliability (performing the service in a dependable manner);
- responsiveness (providing timely and helpful service);
- competence (having the necessary skills and expertise); and
- courtesy (the manner in which customers are treated).

These dimensions are not encountered as objectives in many other processes. No company we have studied has redesigned its processes with the explicit objective of improving courtesy, for example. Yet if these are the service performance criteria that customers care about, they should be a key focus of service companies and companies that supply service along with products.

SERVICE PROCESS INNOVATION STRATEGIES

In services, as in order management, there have evolved several common approaches to transforming business processes. Primary among these is the move toward real-time service delivery. Many existing service processes involve gathering, then analyzing, information about customer needs and characteristics. This is the case, for example, in insurance, bank loans, professional services, and even lodging and transportation services, to the extent that they require some degree of processing and data acquisition before customers can take advantage of them.

Many of these processes are targets in the move toward real-time service provision. Insurance agents with laptop computers can deliver real-time quotes (can real-time underwriting be far off?) and Progressive Insurance offers real-time claims processing, with claims adjusters often issuing checks to insured drivers at the scene of an accident.[20] Citibank offers 15-minute mortgage loans; some hotels are experimenting with processes that allow room access with a credit card and check-out without visiting the registration desk. Only professional services have not moved toward real-time proposal generation.

Another key strategy for service process innovation lies in customer individualization. In a number of service industries, companies that interface with millions of customers each year are creating customer databases that facilitate individualized treatment of customers. Airlines and hotels, for example, that automatically upgrade frequent customers to a higher grade of service build strong customer loyalty. Individualization of service is also found in product businesses. A service representative who knows immediately the details of customers' purchases can make customers who call feel like individuals. With automated number identification, customers are known even before giving their names.

Given the many variables that can affect service quality, one service process strategy is to control or at least monitor as many variables as possible. Federal Express, for example, employs information technology to minimize unexpected factors that might delay package deliveries. A new system for field office managers models and predicts incoming package volumes on the basis of criteria such as day of the week and weather conditions, and the company's aircraft operations function is using a new system for monitoring weather conditions that allows dynamic routing of its aircraft. When mistakes or unexplained factors delay a package, Federal Express's goal is to tell the customer before the customer can tell the company.[21]

A number of companies have innovated in service processes by separating their performance from control and monitoring. General Electric's Appliance Division, Otis Elevator,[22] and Pizza Hut,[23] for example, receive service requests at central national telephone centers that dispatch service from the company location that can best fulfill the request. For roving service providers, portable computing and networking capabilities are available. The vans of GE Appliance service technicians, for example, are equipped with computers and packet radio communication devices. A centralized service coordination approach enables companies to centrally monitor service quality and provides data that can help them better plan new locations; it has even enabled some companies to substantially reduce the size of their field service management staffs.

IT ENABLERS OF INNOVATION IN SERVICE PROCESSES

Although most of the companies that employ IT in service functions have not done so in an explicit process context, their innovations are nonetheless integral to effective processes. A process orientation would probably only increase the value of the IT contribution.

Because many service processes take place in the field, a growing use of IT is to support constant communication with field personnel. This infrastructural step can yield a variety of potential benefits. The ability to communicate immediately the performance of service processes to a central database facilitates the compilation of service and product performance characteristics, and the accessibility of field service representatives linked by a network is a boon to dispatching. A company's communications network can also be used to disseminate information about new products or services, including price changes and promotions. Federal Express,[24] IBM,[25] Motorola, and other companies have used their networks in this manner with great success.

A number of firms have employed IT to build service capabilities into their products, greatly improving the speed and accuracy of their service. Some models of Otis elevators, for example, have devices that constantly analyze their performance and notify service centers when a problem is detected. Some Xerox high-end copiers and publishing systems now offer a "remote interactive communications" capability that alerts Xerox service personnel to impending or actual system failures.[26] Some IBM computers also offer a similar capability.

Expert systems technology has found a valuable niche in services providing problem diagnosis assistance. A number of companies that market technically complex products give customers access to telephone service representatives backed up by expert systems. Westinghouse Electric diagnoses problems in large electric turbines with the aid of a sophisticated expert system at its national service center,[27] and the fountain sales division of a major soft drink producer uses a smaller-scale expert system to diagnose problems with soda dispensers. Such systems allow companies to employ fewer expert service representatives and help them resolve problems faster.

The composite system approach discussed earlier in connection with case management can be applied to service processes as well. A number of firms are trying to consolidate service processes into one role—sometimes referred to as the "help desk."[28] To accomplish this, service case managers must be able to draw from various databases throughout the company. Such systems may include expert diagnostic capabilities.

Quality in service processes also depends heavily on the ability of humans to share their expertise. Service expertise can be captured either in expert systems or in group conferencing systems that provide electronic bulletin boards for sharing problems and ideas. The former technology is illustrated in the expert systems used by financial planners. John Hancock and American Express's IDS division have both developed effective expert systems for use in preparing financial plans, and Barclays Bank, IBM, Digital Equipment Corporation, and Xerox are using electronic conferencing for sharing service expertise.[29]

IT AND INTERORGANIZATIONAL PROCESSES

Many of the processes described above involve the exchange of information across organizational boundaries. Most previous research on these types of systems has focused either on technical issues involving electronic data interchange (EDI) standards[30] or on the potential competitive advantages to be derived from such initiatives.[31] Both are important issues, but neither takes account of the notion that information exchange occurs within the context of processes that cut across organizations.

James McGee, who has focused on precisely this issue in recent research in the consumer products and retail industries, has found the implementation of EDI to involve complex issues related to process integration and the behavior of the individuals who perform the processes.[32] In the absence of careful attention to these issues, processes may suffer from the introduction of technology. At Gillette, for example, where sales representatives had developed short cuts to the formal process, the introduction of EDI actually increased order transmission times.[33] McGee also discovered that interorganizational process strategies in which customers try to dictate the terms of the process and relationship to suppliers do not generally succeed. The process is better viewed as a partnership in which both customer and supplier are expected to compromise and adjust their behavior.

A process orientation to interorganizational systems presumes that processes are designed and described before systems are implemented. The technologies for interorganizational systems must be agreed upon by all parties—ideally at an industry level, so that multiple proprietary infrastructures do not have to be built. And because customer-oriented processes often eventually end up strongly influencing internal processes such as financial and marketing activities, these should not be designed without taking interorganizational processes into account.

McGee contends that taking a process approach to interorganizational systems will greatly increase the value of the investment to the involved parties. The greater the integration between buyers' and sellers' processes, the greater the technology's value. Such integration can only be achieved, however, through joint design of the interorganizational process by all involved organizations, including third-party suppliers of EDI services.

SUMMARY

Processes at the customer interface are perhaps the most critical to an organization's success. They are essential to a firm's cash flow, customer satisfaction, and process efficiency. Because they often involve individuals who resist structure, these processes have not received much attention in the past. Yet opportunities for innovation in customer-facing processes, in marketing and sales management, order management, and service processes, are manifold.

Key enablers of innovation in these types of processes can be found in both IT and human resources. Interorganizational networks can increase speed and responsiveness; expert systems support the distribution of expertise to customers and the field. Empowering frontline employees is critical to process success for almost any new design, and case management combines multiple enablers to create designs that represent entirely new organizational forms.

In the next chapter, we consider processes even less structured than those at the customer interface—management processes.

Notes

[1] For a detailed discussion of closed-loop marketing processes, see Robert C. Blattberg and John Deighton, "Interactive Marketing: Exploiting the Age of Addressability," *Sloan Management Review* (Fall 1991): 5–14.

[2] John D. Little, "Automated Newsfinding in Marketing," *MIT Management* (Winter 1990): 26–31.

[3] Benn Konsynski and Michael Vitale, "Baxter Healthcare Corporation: ASAP Express," 9-188-080. Boston: Harvard Business School, 1988, rev. 1991.

[4] Jerome D. Colonna with Stephen Sabatini, "Keeping American President Afloat," *InformationWeek* (August 15, 1988): 28–32.

[5] Ramon O'Callaghan and Michael Vitale, "American Crown," 9-188-052. Boston: Harvard Business School, 1988, rev. 1989.

[6] John J. Sviokla, "An Examination of the Impact of Expert Systems on the Firm: The Case of XCON," *MIS Quarterly* 14:2 (June 1990): 127–140.

[7] For a discussion of Digital's implementation efforts on XSEL, see Dorothy Leonard-Barton, "The Case for Integrative Innovation: An Expert System at Digital," *Sloan Management Review* (Fall 1987): 7–19. For a more recent analysis of sales-force configuration-system implementation for which the company is not identified, see Mark Keil, "Managing MIS Implementation: Identifying and Removing Barriers to Use," Ph.D. diss., Harvard Business School, 1991.

[8] J. Hammond and C. Couch, "Digital Equipment Corp.: Complex Order Fulfillment," 9-690-081. Boston: Harvard Business School, 1991.

[9] Melissa Mead and Jane Linder, "Frito-Lay: A Strategic Transition (A) & (B)," 9-187-065 and 9-187-123. Boston: Harvard Business School, (A) 1986 and (B) 1987; also Nicole Wishart and Lynda Applegate, "Frito-Lay, Inc.: HHC Project Follow-Up," 9-190-191. Boston: Harvard Business School, 1990.

[10] Max D. Hopper, "Rattling SABRE—New Ways to Compete on Information,"

Harvard Business Review (May–June 1990): 118–125; also, for a recent mention of American's process innovation efforts, see Glenn Rifkin, "Max Hopper: 'The Big Systems You Can Use to Leap Ahead of the Competition Are Becoming Increasingly Impossible to Achieve'," *ComputerWorld* (August 5, 1991): 65.

[11]John J. Donovan, "Beyond Chief Information Officer to Network Manager," *Harvard Business Review* (September–October 1988): 134–140.

[12]Marie-Christine Monnoyer and Jean Philippe, "Using Minitel to Enrich the Service," in Ewan Sutherland and Yves Morieux, eds., *Business Strategy and Information Technology* (London: Routledge, 1991): 175–185.

[13]Stuart Madnick and R. Wang, "Evolution Towards Strategic Applications of Data Bases Through Composite Information Systems," *Journal of MIS* 5:2 (Fall 1988): 5–22.

[14]See, for example, Jan Carlzon, *Moments of Truth* (New York: HarperCollins, 1989); also Thomas J. Peters, *Thriving on Chaos* (New York: Harper & Row, 1987).

[15]Adam Smith, *The Wealth of Nations* (Edinburgh: Adam and Charles Black, 1850): 3.

[16]J. Richard Hackman and Greg R. Oldham, *Work Redesign* (Reading, Mass.: Addison-Wesley, 1980).

[17]James D. Thompson, *Organizations in Action* (New York: McGraw-Hill, 1967).

[18]Robert D. Buzzell and Bradley T. Gale, *The PIMS Principles* (New York: Free Press, 1987).

[19]Valarie A. Zeithaml, A. Parasuraman, and Leonard L. Berry, *Delivering Quality Service* (New York: Free Press, 1990).

[20]"The Right Mix of Skills," *InformationWeek* (March 4, 1991): 12.

[21]Personal interviews with Federal Express executives in September 1990.

[22]Michael Vitale and Donna Stoddard, "Otisline," 9-186-304. Boston: Harvard Business School, 1986.

[23]"Mr. Winchester Orders a Pizza," *Fortune* (November 14, 1986): 134.

[24]For information on Federal Express' field technologies, see David H. Freedman, "Redefining an Industry Through Integrated Automation," *Infosystems* (May 1985): 26–32; also Arilyn Stoll, "For On-The-Road Firms, Hand-Held Terminals are Pivotal," *PC Week* (August 22, 1988): c11–c12.

[25]Lynda M. Applegate and H. Smith, "IBM Computer Conferencing," 9-188-039. Boston: Harvard Business School, 1990.

[26]Personal interviews with Xerox executives, May 1991.

[27]R. Jaikumar, "Westinghouse Steam Turbine Generator Diagnostic System (A)–(C)," 9-686-006, 9-686-009, and 9-686-059. Boston: Harvard Business School, 1986.

[28]Interview with Dennis Yablonsky of Carnegie Group, reported in "Artificial Intelligence: Brain Teaser," *The Economist* (August 17, 1991): 63.

[29]Many of these information-sharing technologies are described by Thomas A. Stewart, "Brainpower: How Intellectual Capital Is Becoming America's Most Valuable Asset," *Fortune* (June 3, 1991): 44–60. See also Applegate and Smith, "IBM Computer Conferencing."

[30]See, American National Standards Institute, "An Introduction to Electronic Data Interchange," July 1987.

[31]See, for example, James I. Cash and Benn Konsynski, "IS Redraws Competitive Boundaries," *Harvard Business Review* (March–April 1985): 134–142.

[32]James V. McGee, "Implementing Systems Across Boundaries: Dynamics of Information Technology and Integration," Ph.D. diss., Harvard Business School, 1991.

[33]James V. McGee, James I. Cash, and Benn R. Konsynski, "The Gillette Company (A)–(D)," 9-186-158 and 159, and 9-188-011 and 012. Boston: Harvard Business School, 1986.

Chapter 13

Management Processes

Of all the processes in an organization, management processes are the most poorly defined, and least likely to be viewed in process terms. Indeed, some would argue that the term "management processes" is an oxymoron, at least insofar as we define process in this book. The structure of management activities is rarely documented, and such activities often are not performed in the service of specific customers. Nevertheless, we believe that the concept of process innovation is of value to the work of managers. Essentially, we are arguing for a greater degree of structure in managerial work. Managers unwilling to accept more structure in the form of a process orientation will be unable to take advantage of information technology and other change enablers to become more efficient and effective.

Management processes involve planning, direction setting, monitoring, decision making, and communicating with respect to a firm's key operational processes and assets. Given this definition, examples of management processes in firms include:

- strategy formulation,
- planning and budgeting,
- performance measurement and reporting,
- resource allocation,
- human resource management,
- infrastructure building, and
- stakeholder communications.

These are not the only activities of managers, but they are the most structured and thus the most likely to be conducted within a process setting. More interpersonal processes, including leadership and influence building, are extremely important to managerial success, but may be outside the realm of a process orientation.

Thus far, we have focused on operational processes—processes related to how products or services are created, produced, sold, or serviced. This emphasis is reasonable given that operational processes are where money is made. But management processes are where much of the money is spent. In the United States, for example, "executive, administrative, and managerial occupations" comprised 13% of the full-time work force in 1990, and was one of the highest-paid of any major category of worker.[1] Administrative workers, who presumably implement management processes and decisions, made up another 18%. Together these two categories were larger than any other. For this reason alone, management processes are worthy of attention.

Management processes warrant attention as separate processes. Because there are management aspects to the operational processes of organizations, some firms have elected not to address them separately. Of IBM's original 15 major processes, for example, only "financial analysis" might be considered management, and it is related primarily to pricing. Three processes IBM subsequently added—human resource management, accounting, and information technology infrastructure—have management aspects, but are more involved with administration than strategic decision making.

Firms that do not address management processes separately run two risks. One, if management tasks are not within the normal flow of work activities, they may not receive sufficient attention. And two, without a separate focus on management processes, the management activities identified in operational process work may not be consistent across different processes. Given the importance of management and the need to integrate management and operational activities, it is reasonable in most cases to consider the management process both separately and as a set of activities within operational domains.

Although a number of firms have begun to focus explicitly on management processes, it is too early to know the results of their initiatives. The exploration division of British Petroleum, for example, has designated the management reporting and performance measurement process one of the first to be redesigned. Similarly, British Telecom has selected its planning and budgeting process as a candidate for early analysis; Aetna and Capital Holding, two large insurance companies, are beginning to redesign their financial and reporting processes. At Xerox, the entire "business management" process was identified as one of 14 key processes to be redesigned. And at Rank Xerox U.K., the human resource management process was among the first to be redesigned and, because of its importance, was the first process selected for systems implementation.

Given the early stage of all these efforts, our belief in the importance of process innovation in management processes is necessarily based more on faith and intuition than on hard evidence. We have observed shifts in the organizational location of management activities—from centralized to decentralized, and vice versa—that have made little difference in management effectiveness because the processes were not fundamentally changed. We have seen many attempts to apply information and information technology to the work of management fail in the absence of a management process. Finally, we have seen the benefits of applying process thinking to other types of processes that involve work as unstructured as that of managers. For all these reasons, we believe that management process innovation is worth pursuing.

MANAGEMENT PROCESS AND PRACTICE

Despite many years of formal and informal discussions about management, only in the past two decades have we seen empirical research concerned with what managers really do. The most useful of this research for our purposes comes from Henry Mintzberg, John Kotter, and Daniel Isenberg. Their conclusions, after observing senior managers at their work, are of great relevance to our deliberations on management process.[2]

Managerial work, according to Mintzberg (and others before and after him), is highly unstructured and discontinuous. He characterizes management as involving three primary roles—interpersonal, informational, and decisional. Managers, in his view, devote little time to any single task and spend much of the day responding to ad hoc events. Managers do have regular duties, some associated with ceremonial functions and interaction with customers, and they process a great deal of both factual and impressionistic information communicated through many different media. Although some managers rely on information technology for a small part of their information flow, managerial activities for the most part have not changed substantially for more than a century.[3]

Kotter, who has heavily researched managerial behavior, takes a process view of managerial activity, albeit not as results-oriented a view of processes as we take in this book. In Kotter's view, the key management processes are those that involve agenda setting and network building.[4] Agenda setting is the process of creating loosely structured organizational goals, plans, and strategies. Although formal planning processes are part of agenda setting, they include many less structured aspects. Network building, according to Kotter, is a process by which managers establish and maintain communication linkages with their relevant worlds, both inside and outside their organizations. Nodes on these networks are any sources of information that can help advance the executives' agendas.

Although these two processes are important to executives as individuals, from the perspective of the entire organization there are other equally important management processes. Moreover, it is not clear that breaking managerial work into these two processes is likely to convince managers to structure those processes better. Agenda setting and network building are highly individual and interpersonal processes, and managers do not necessarily describe or measure them uniformly or share common assumptions about their improvement. In other words, Kotter's process model is helpful from a descriptive, but not a prescriptive, perspective.

More recent research has addressed the issue of how managers think. Isenberg, who has been a primary advocate of a more cognitive approach to management behavior, came to the following conclusions.

> It is hard to pinpoint if or when they [senior managers] actually make decisions about major business or organizational issues on their own. And second, they seldom think in ways that one might simplistically view as "rational," i.e., they rarely systematically formulate goals, assess their worth, evaluate the probabilities of alternative ways of reaching them, and choose the path that maximizes expected return. Rather, managers frequently bypass rigorous, analytical planning altogether, particularly when they face difficult, novel, or extremely entangled problems.[5]

Although it is undeniably useful to know the facts about managerial thinking and behavior, these authors are not altogether sanguine about their findings. They see several problematic aspects to management practice. Management sometimes seems to be a superficial activity, never getting to the core of an issue before moving on to the next one. Managerial information is too abundant and at the same time too unstructured for it to be shared throughout the organization. In terms of time management and productivity, management behavior is inefficient, and in terms of its standard of evidence and analysis, management decision making is irrational.

How are we to deal with these findings in the context of process innovation? In the vast majority of organizations, management is the one area, more than any other, that lacks a well-defined sense of process. There is no structured set of activities that lead to a stated outcome, no customer for that outcome, and no measure of how the outcome is produced. As a result, we typically never know how efficient or effective the activities of management really are. Clearly, the installation of any process, however well described or structured, would be a step forward.

Yet given the findings of Mintzberg, Kotter, and Isenberg, we are well advised to exercise caution in our advocacy of greater structure in management processes. Any process designed for managers must accommodate intuition, creativity, exceptions, and serendipity—because they will occur whether designed in or not, and because they are often of great value. We believe it is possible to design processes that accommodate these positive aspects of management behavior; the real issue is whether managers will practice them or not.

The quality movement, particularly the Japanese variant, has yielded some insights into management processes. According to Imai, the application of quality thinking to management involves two distinct approaches (see Figure 13-1).[6] One of these is cross-functional management, wherein managers focus not only on managing their own function, but also on managing for specific process objectives across functions. Imai reports that this type of management has been practiced by Toyota since 1962. Although discovered here more recently, cross-functional management processes are well known to Western firms embarking upon process innovation initiatives.

Figure 13-1   The Role of Quality in Management

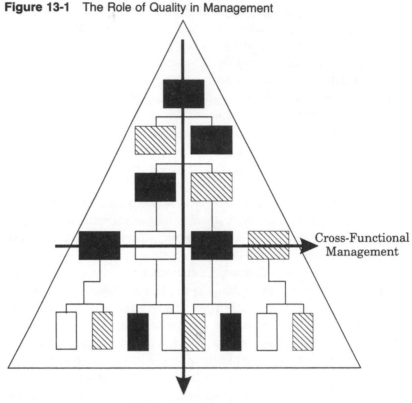

Cross-Functional
Management

Policy Deployment

The second aspect of quality in management, often called
policy deployment, or *hoshin kanri* in Japanese, relates to the
alignment of policy, or strategic goals, through all levels of the
management hierarchy. With effective policy deployment, all levels of workers understand their firm's key strategies and their
role in implementing them. To achieve this objective, management goals must be clearly defined in terms that are observable
and measurable, and the measures, both of strategic results and
of the processes designed to achieve them, must be clearly understood and applied at regular intervals.[7]

In most firms, the implementation vehicle for policy deployment is the policy deployment form, which every worker must
complete at regular intervals. On it, all levels of strategy and the
tactics for implementing them are recorded, down to the level of

the worker completing the form. This enables functional managers to readily assess, for example, whether their functions are supporting overall divisional or corporate goals, and plant managers to know the specific objectives they are responsible for achieving. Nonmanagerial workers develop with their supervisors a personal agenda for meeting the objectives of the supervisor's unit. Through this medium, each employee comes to understand the cascading objectives of the firm and the meaning of his or her own job relative to those objectives.

This approach is beginning to be implemented in the West. Rank Xerox U.K., for example, has developed a classical version of a policy deployment initiative that establishes a set of common goals that address, in the company's terms, "what we will do, what we need to do it, and how we will do it." Each manager and employee translates, with management assistance, these company goals into specific personal goals. It is too early to see financial results from the initiative, but the firm's employee satisfaction percentages are up.

Rank Xerox U.K. has also created a well-defined process for reporting and acting on firm performance. The stated goals of the process are:

- to shape behavior by clearly defining meeting forums and frequencies, and

- to shape managerial thought by specifying reporting formats and decisions to be made.

In practice, this process entails producing, for each management meeting, a one-page statement of purpose and four required reports, as well as explicitly identifying the chairperson, attendees, meeting manager, day of the month, and prerequisite events. Reports are generally in graphic format and address such matters as the previous month's results, root causes of the business performance shown, and actions to be taken. Since the process was initiated in 1991, senior management has observed a radical decrease in the time devoted to meetings each month (from approximately eight days to one) and much greater focus in meetings on issues of real importance to the business. With the purpose of a meeting and format in which information is to be presented determined in advance, there is more time to address the true problems and causes of business performance.

MANAGEMENT PROCESS INNOVATION

The activities discussed above are necessary steps toward, but are not in themselves, management process innovation. Like quality-oriented approaches to manufacturing, they emphasize minimization of variation in strategic and tactical goals. They do not involve striving for breakthroughs in managerial performance (whatever that would mean) and do not employ specific technological or human enablers to transform management behavior.

To seriously advocate process innovation in the management domain, we must first discuss some likely objectives and attributes of management processes. Like any other process, management process redesign (or, more likely, design) is most likely to be effective, as is most any process redesign effort, when there is a vision for the process. The attributes of process vision described earlier are relevant here. Vague notions of managerial efficiency and ill-defined goals of providing managers better information for decision making are unlikely to lead to focused managerial process innovation. For managerial processes to be the target of innovation, someone—presumably the CEO—must have a clear view of what is wrong with, and how to remedy, existing approaches to management.

The types of dissatisfaction that can lead to process visions can arise both from the process of management (the time it takes, the number of people involved, specific behaviors) and from its result (the decisions made by managers, the outcome of those decisions, or the quality of other management outputs, including plans, initiatives, subordinates' careers, and so forth). Process-oriented dissatisfaction should focus not on narrow goals of improving the efficiency of management activity, but on criteria that affect performance such as speed and value-added to the line organization.

Speed in decision making is beginning to emerge as a real issue in the managerial processes of some companies, although it is important to point out that we still have no clear understanding of the implications of slow decision making. When speed becomes a primary objective of such processes as product development, order management, and customer service, it is logical that it become a concern in management activities. As Kathleen Eisenhardt, the primary investigator of this phenomenon, has noted: "Strategy making has changed. The carefully conducted industry analysis or the broad-ranging strategic plan is no longer a guarantee of success. The premium now is on moving fast and keeping pace. More than ever before, the best strategies are irrelevant if they take too long to formulate."[8]

Eisenhardt discerned five distinguishing characteristics in high-technology managers who were considered to be rapid decision makers. She observed that they track their businesses and industries in real time, consider multiple alternatives in parallel, consult only with a few trusted advisers, seek consensus but act on their own if they don't achieve it, and fit individual decisions into an overall collection of decisions and plans (similar to Kotter's agenda-building process). Executives who exhibit these characteristics typically take only two months to make decisions that take other executives as much as a year.

Another aspect of management process (as opposed to result) is the element of learning incorporated in management behavior. A growing chorus of managers and researchers is arguing the benefits of a learning-driven approach to management. Although this approach is more philosophy than well-defined objective, managers are integrating it into their activities to varying degrees.[9] Among the hallmarks of learning-oriented management are: structuring time for thinking into the management time budget; employing "systems thinking" (the most recent version of Forrester's "systems dynamics" approach) to reveal underlying patterns in business activities; and treating planning processes as learning exercises (rather than assuming that they yield truth).[10]

Another possible management process vision is that of managers acting on fact rather than on intuition. The findings of Mintzberg and Isenberg suggest that this would be an uphill struggle, yet a number of CEOs we interviewed in the past year expressed a strong conviction of the need for fact-based management. Some have linked the search for facts with the attempt to bring quality to management, in a manner reminiscent of policy deployment and the Rank Xerox example. Few of these have progressed very far in understanding what kinds of facts are desirable for what purposes.

An objective for management process results that we have

seldom encountered in companies is a vision of better management decisions resulting from more factual management information, learning, and so forth. Few firms seem to have the nerve or ambition to assess decision quality systematically; they seem to be afraid of discovering failure. Yet failure is a key component of learning. Many physicians regularly participate in forums in which they examine failures in treating or healing patients. Perhaps because failure is not as inevitable in management as in medicine, it is considered anathema in most corporations. To study the results of decisions would inevitably turn up failures, so we never learn how often we fail, examine why we failed, and learn from the failure.

Improved decision making as part of a management process vision also relies on improvements in controlling and reporting processes. This is the objective, often implicit, of many initiatives to improve management information. The specific goal may be to get information to managers faster so they can act faster, to achieve greater local or centralized control, or to provide different types of information to managers. We have seen sporadic efforts over the years to integrate nonfinancial measures of corporate performance into the management control process. Although logic suggests that such measures can provide warnings of problems with or raise issues related to corporate performance earlier, few firms have successfully identified and then integrated nonfinancial measures into their management agendas.[11]

Finally, management process objectives can target the improvement of specific functions or, better yet, operational processes. Frito-Lay's management-process efforts, for example, were directed primarily at better managing marketing and sales processes. A large pharmaceutical firm we studied focused its managers' attention on the order fulfillment process, whereas Cincinnati Bell's processes and executive information systems were directed at improving levels of customer service.[12] A computer company studied by Rockart and DeLong was interested primarily in converting its managers from a technology to a marketing focus.[13] Given the time it takes to transform a process, whether management or operationally oriented, the specific domain for the process vision should be broad enough to be of long-term interest. In some firms, a key subprocess of the strategy formulation process might involve deciding what the key targets of management processes should be.

A clear vision of what is wrong with and how to fix existing management processes, then, is a necessary prerequisite for management process innovation. The vision may be of either the management process or its result, but it must be clear, measurable, and inspirational enough to motivate behavior change; because managers will have to change substantially before a process orientation can take hold, the vision must be even more compelling than for most operational processes. Management process innovation can be furthered by the enablers discussed elsewhere in this book. We now turn to the strategies for establishing and changing specific management processes, with a nod to their respective enablers.

KEY MANAGEMENT PROCESSES AND ENABLERS

The processes that senior management is responsible for can vary across firms, but there is a common set. The processes described below are almost invariably consigned to senior management, and they appear in virtually every industry. For each process, key directions for innovation are described, together with specific enablers (listed in Figure 13-2).

Strategy formulation and strategic decision making.
Among unstructured management processes, strategy formulation and strategic decision making are the least structured. Although there are formal aspects to this type of process, such as preparation of strategy documents, they correlate only partially with the strategic agendas of executives. Yet even strategic activities within this process area—for example, considering a merger, restructuring, divestiture, or new product line—can be extremely time- and resource-consuming for senior management. Furthermore, while strategic decisions hang in the balance, the focus and productivity of lower-level employees is often impaired.

What can a "designer" of the strategy process do to improve it? Given the unstructured state of the existing process, and the potential resistance of senior executives to process innovation, a minimalist approach—perhaps process improvement rather than process innovation—is probably best. The improvement process must be highly participative, and the actions to be taken must fall into the categories of description, measurement, information, and external linkages.

Figure 13-2   Enablers of Innovation in Management Processes

- Executive information systems that provide real-time information
- Electronic linkages to external partners in strategic processes
- Computer-based simulations that support learning-oriented planning
- Electronic conferencing and group decision-support systems
- Expert systems for planning and capital allocation
- Standard technology infrastructure for communication and group work
- Standard reporting structures and information
- Acknowledgment and understanding of current management behavior as a process
- Accountability for management process measurement and performance

A high-level internal or external consultant could describe the process followed in several recent strategic initiatives or strategy formulation exercises, and then compute a rough set of time and cost measurements. Rather than make recommendations for new processes, the consultant should simply reveal the results of the analysis in a nonjudgmental fashion. Senior managers permitted to observe the steps and resources consumed by a strategic process may begin themselves to change their behavior.

The steps to be taken in terms of information and IT are also modest for strategic processes. As observers of management behavior have frequently noted, many executives actively resist the use of computerized systems created to support their work.[14] Eisenhardt's recommendation that decision making be speeded up through the provision of real-time information obviously has implications for electronic information systems, although if the information is simply to be read rather than analyzed and modeled, paper may well suffice. Electronic information delivery should be considered only for executives who want access to a broad range of up-to-the-minute information and a capacity for "deep drilling" when desired. An executive culture that does not frown on keyboard use is also prerequisite for electronic information delivery with current technology.

For the headquarters organization of Imperial Chemical (ICI) in London, an executive information system was developed with the sole objective of providing real-time information. According to the manager responsible for its implementation, the system was well defined and focused on its intended purpose (although, as noted above, paper might have been sufficient).

> We have no product of our own, and no customers of our own. We don't have any of our own business data. The only data that we have on ICI's business is what is sent to us on a statutory basis . . . Therefore we have no need whatsoever for the classical drill-down techniques. The role of headquarters is to set the strategy for the ICI group over the next ten years, what shape we want to be in ten years' time and how we get there. What we have done is developed an EIS to support that objective.[15]

Information can also be exchanged electronically with partners in strategic processes. Interactions with external legal, consulting, or investment banking firms are often a major component, and constitute a major proportion of the cost, of strategic processes. A few firms, among them Xerox and General Motors, have mandated electronic communication in standard formats with their external law firms.[16] Xerox also communicates electronically with its public relations firm. Mandates could be issued to other types of professional service firms. This approach, although incremental, can reduce both the time and cost of working with external strategic process partners, and constitutes a step toward a more structured relationship that works along process lines.

If empowerment is the primary human resource enabler for operational processes, accountability takes its place at the senior management level. Senior-level managers may be wholly empowered, but unwilling to accept responsibility for the strategic process, and they may resist measurement of the time and cost of the process and the tracking of results of decisions. Accountability is essential if strategic processes are to be described and become targets of innovation.

Planning and budgeting. Other major, albeit more tactical, processes that lie in the purview of senior management are planning and budgeting. Because they are more structured than the strategy process, these processes offer greater opportunities for process innovation. Beyond time and cost reduction objectives lie significant opportunities for improving level of participation and fostering iteration and learning.

Planning and budgeting processes are notorious for their rigidity and irrelevance to management action. Rigid adherence to a process of rapid or efficient completion may only make the process less relevant to the true management agenda identified by Kotter.

Managers have recourse to a number of different approaches for making planning and budgeting more relevant. The use of scenario and simulation techniques, for example, can help managers think about the consequences of various actions. Senge and de Geus describe a shift in the planning processes at Royal Dutch/ Shell from "producing a documented view of the future" to "design[ing] scenarios so that managers would question their own model of reality and change it when necessary,"[17] a shift that was rewarded handsomely by the company's successful actions during the oil shocks of the 1970s and 1980s.

Computer-based simulation tools can be a vehicle for learning-oriented planning, although their most impressive use thus far has been retrospective. The Stella tool for dynamically simulating the many variables that affect company performance, for example, is an excellent teaching tool, although its most powerful use is for retrospective analysis of performance problems. Stella simulation of the People Express business is highly insightful about the reasons for the firm's failure.[18] But although it is possible to do so, we know of no firm that has used Stella extensively to better understand the future rather than the past.

Information technology can also be used to increase participation, and/or the number of iterations (and thus the potential for learning) in planning and budgeting processes. Electronic conferencing and group decision-support systems could be used to involve multiple individuals in scenario-based planning, or simply to receive inputs about planning and budgeting issues. Electronic transmission of planning documents can support (as it has at Colgate-Palmolive, which uses its network to plan on a worldwide basis[19]) multiple iterations of plans or budgets during the same planning cycle.

Expert systems are a potentially valuable technology for improving the accuracy of planning and budgeting processes, although we know of no implementations in these processes. A computer firm's somewhat futuristic video, however, illustrates the use of expert technology to lead a manager through the annual budget preparation process.[20] The system's knowledge of past budget inputs and characteristics of the unit being budgeted make it a valuable assistant.

Performance measurement and reporting. Measurement and reporting processes are in place and well established in most every company, largely because of external (e.g., SEC) reporting requirements. These processes draw heavily on resources of both managerial and administrative personnel. At the highest levels of the firm, process innovation should focus not only on the delivery of performance information, but also on its use. For administrative purposes, the process goal is to improve the quality, speed, and accuracy of the information supplied in the most efficient manner possible.

Most companies we have studied want improvement rather than innovation in their measurement and reporting processes. They want these processes to use additional measures, be more consistent across the firm, or be less burdensome for the line managers who develop performance reports. But because processes for corporate and human-resource performance assessment tend to be linked, process change initiatives in this area tend quickly to become major efforts.

From an administrative perspective, the greatest reporting process benefits probably derive from the use of standard data and applications for reporting. When all financial data elements are consistent and every business unit around the world uses the same applications for generating and analyzing them, reporting processes and the business rules within them can begin to be described and simplified. This is a major focus at Dow Chemical, which has selected an integrated financial-management package that it intends to deploy worldwide. Afterward, the company will begin to supply business units with standard business rules and processes. The controller believes that establishing common business practices and reducing the number of systems interfaces will permit radical simplification of the firm's financial and reporting processes.[21]

Standard reporting processes are difficult to establish in global, diversified firms such as Dun & Bradstreet. This global, diversified service firm found that it needed different measures and

reporting structures for different types of businesses and geographical units, but as measures and reports proliferated they
tended to be requested broadly throughout the company. When
business units complained about the reporting burden, Dun &
Bradstreet corporate financial managers mounted an initiative
to simplify reporting processes and pare down measures to those
essential to running the business. Xerox, which similarly decided
to simplify reporting structures several years ago, found it difficult to require the same information of all its diverse businesses,
and has reverted to a more heterogeneous set of measures and
reporting processes.[22]

At senior management levels, the greatest process benefits
in the reporting and measurement area come from relaxing some
of the assumptions that limit the process's effectiveness. For example, most reporting processes assume that financial indicators
of performance are the most useful, and they generate and analyze
such measures at regular reporting intervals (e.g., quarters or
months). But as Robert Eccles has pointed out, the most useful
information may involve quality, customer service, or product
pipeline metrics, and the relevant reporting interval often changes
as business accelerates or decelerates.[23] Moreover, decisions must
be made within intervals, and real-time information would be
extremely useful in making them.

Some of the most useful reporting may involve soft indicators
such as measures of corporate morale and customer satisfaction.
Survey results can be reported on-line, of course, but a more
interesting set of indicators may be how employees talk to one
another. IBM employees' reactions to John Akers' internal speech,
widely disseminated over electronic mail and in the press, were
captured in part by an internal electronic conferencing system.[24]
Although they may not have relished reading the many negative
reactions to Akers' remarks, IBM senior executives were afforded
a unique opportunity to take the pulse of the organization.

Better reporting is cited as the primary objective of many
executive information systems. Rockart and DeLong, in fact, cite
improved planning and control as the key application and most
significant impact of executive support systems (as they refer to
the technology). They identify six major benefits of using executive information systems to improve, if not innovate, measurement and control processes:

- improvements to existing corporate or divisional systems (e.g., improved data collection, integrity, or reporting formats);

- accommodation of new or different management roles in redesigned reporting processes;

- the ability to perform ad hoc analyses;

- enhancement of personal communication links;

- improved program and project management; and

- changes in planning and forecasting processes.

Because none of these benefits has been evaluated in a formal, measured, process context, it is difficult to know to what degree reporting and control processes have benefited from EIS technology. Perhaps as a result of the absence of process thinking about management reporting and strategic decision making, executive information systems have not been entirely successful in transforming management behavior. Failure rates for such systems are estimated at between 50% and 70%.[25]

Executive information systems are also commonly used to improve reporting for operational processes. Particularly in information-driven businesses, executives can use better reporting technologies to intervene actively in a process when they see the need or opportunity to do so. A frequently cited example occurred at Phillips Petroleum, where now-retired division head Robert Wallace used an EIS to actively monitor the operational process of crude oil trading.[26] Wallace's description points up both the value of and problems associated with such use.

> We made $50 million last year on crude oil trading. We made that driving things very much from the executive level. We had managers and traders and all of the backroom people, but the real dollars were made because senior management was staying on top of all the events we could perceive on a daily basis that were moving us towards some opportunities. And we were able, through our increased understanding, to quickly make a decision to act. The alternative is to sit in an ivory tower waiting for the guy down below to work up enough courage, and perhaps insight, to start pushing his notion of a business opportunity up the system. In that case, by the time the idea gets to the top the chance may be gone.[27]

It is difficult to argue with Wallace's results, but one can certainly dispute his methods and conclusions. The "managers and traders and all of the backroom people" must have felt thoroughly disempowered by Wallace's intervention. Had he allowed the traders to act quickly on information themselves, rather than have to push their ideas slowly up through the system, the process might have worked better. The disempowering effect of EISs has been seen at other companies; Xerox senior managers, for example, were forbidden to delve down more than two organizational levels into operating results for fear that lower-level managers would be disempowered.[28]

Resource allocation. Allocating scarce resources across the organization is a major responsibility of senior management. The capital allocation process is perhaps most familiar, but other resources—talented people, research capability, technology, and so forth—may also be scarce.

Inasmuch as capital allocation may be the only process-oriented management activity in many organizations, simply including other types of resources in the allocation process is one form of innovation. At Johnson & Johnson, for example, where research and clinical trials capabilities in the drug development process were the scarce resource, a tracking system for allocating resources to development projects across J&J divisions became the vehicle for improving the process.

Although most capital allocation processes are well defined, the outputs of the process are frequently overridden by senior management, whose interventions, if they do not yield poor results, may yield poor morale. In firms in which the process is well defined, but some people are much more familiar with it than others, disproportionate allocations of capital may go to the knowledgeable.

To address these issues, Texas Instruments developed an expert system to improve the capital allocation process.[29] Managers in a fast-growing and capital-intensive TI division were concerned that the time and experience necessary to prepare capital budget request packages would become an obstacle to growth. The packages were complex and time-consuming to complete, and few employees had the requisite knowledge to prepare them accurately. The resulting expert system has greatly improved the request process, enabling packages to be prepared in far less time than required by the manual approach and to conform more closely to company guidelines. One experienced employee reported a reduction in preparation time from nine hours to 40 minutes.

Westinghouse developed a system to support not the requestors, but the allocators, of capital.[30] Managers who have used the system, which employs group decision-support technology, report that it makes the allocation process not only more efficient, but also less biased in favor of requestors known to individual executives.

Human resource management. Perhaps because of the bureaucratic nature of the human resource function in many firms, there are well-established processes for many aspects of personnel management, such as career planning, individual performance evaluation, skill building, and so forth. Yet few firms are pleased with their human resource management processes.

There are several barriers to improving human resource processes. One is lack of commitment by senior executives. The frequency of the assertion in annual reports that "people are our most important asset" is in no way matched by the frequency with which managers act as if they were. A second problem with human resource processes is inattention to results. For example, although most managers and supervisors diligently perform evaluations and reviews of their employees, they are rarely given an overall scorecard on the effectiveness and growth of their people. Emphasis on human resource management in the quality movement is beginning to lead to greater measurement and results reporting, but only in quality-driven firms.

Nevertheless, some organizations are placing innovation in human resource processes at a high level of importance. Among the strategies being employed are more focused human-resource reporting, competence-based process philosophies, and moving process responsibility into the field. The process goal of a major pharmaceutical firm is to begin tracking not all human resources, but only outstanding performers whose loss would greatly affect the company. As the firm's chief financial officer puts it, "I get reports every month on how many people we have gained and lost in the research function. Yet I can't tell whether they are Nobel Prize winners or test tube washers." This concern is to some degree a function of scale; another CFO at a successful but much smaller genetic engineering firm sees key researchers every day at lunch and so needs no formal process to evaluate their worth.

At Bass plc in the United Kingdom, the corporate heads of personnel and information technology have formed a partnership to analyze processes with respect to the competence of their performers. Specific competencies, they believe, will be both the greatest enablers of and barriers to process innovation. If successful, their approach, now at an early stage, will be a good means of tying human resource processes to other operational and management processes within the firm.

Bristol-Myers Squibb elected to move human resource processes into the field. Prior to the installation of a common, corporate human-resource system, most of the company's human resource activities had been the province of functional managers or employees, either at the corporate level or in the divisions. When the new system was installed, the functional managers realized that many of the required subprocesses could be performed by line managers and employees in the business units. The common system enabled the use of common, well-defined business rules within the human resource process.

Stakeholder communications.   Senior management is generally responsible for determining with what parts of the outside world—including employees, investors, Wall Street analysts, governmental bodies, residents of the community surrounding the firm, and so forth—the company will communicate. Although we lack experience in redesigning this communication process, conversations with representatives of the different stakeholder communities have convinced us that change is desirable.

With respect to communicating firm performance to investment advisers, for example, neither side seems to have a good understanding of what is possible or desirable from the other's perspective. Senior managers, who view analysts as being concerned only with short-term financial results, do not bother to develop or report other types of measures. Analysts, on the other hand, profess to want better, longer-term, nonfinancially oriented information on corporate performance, but believe that senior management teams either do not have such information or do not want to reveal it to outsiders. We suspect that a systematic dialogue between these groups would be enlightening to both.

The issues are similar for other relationships. It is rare for two parties to a communication to systematically address the requirements for information and possibilities for what can be communicated. To do so would be a major step forward. As for the strategy formulation process, establishing a structured set of activities related to stakeholder communications would constitute a process innovation.

Infrastructure-building processes. The building and maintenance of infrastructures, both human and technical, can also be viewed as a process in the senior management domain. Bristol-Myers's human resource system is an example of a process strategy that is more infrastructure than initiative, having been planned as a platform for a future set of benefits rather than to meet a short-term need. Most attempts to improve or innovate processes involve specific initiatives. It is much easier to get funding, for example, to build a new plant or introduce a new advertising campaign than to build employee leadership skills or a high-bandwidth communication network.

The preference for initiatives over infrastructure is seen all the time in the information systems function, which can readily get funding to build specific systems that meet user needs, but has a much more difficult time (so difficult that it rarely happens) obtaining funding to improve the overall effectiveness of the systems-building function. As a result, investments that could lower the costs of building each system are never made.

Some firms are good at infrastructure-building processes. American Airlines recently made major investments in both skills and technology infrastructures with its "Committing to Leadership" management development program and InterAAct technology platform for communications and effectiveness-building applications, two initiatives designed to be mutually reinforcing.[31] The goal of InterAAct, according to Max Hopper, the firm's head of information systems, is

> not to develop stand-alone applications but to create a technology platform—an electronic nervous system, capable of supporting a vast array of applications, most of which we have not foreseen. InterAAct is an organizational resource that individuals and groups can use to build new systems and procedures to do their jobs smarter, better, and more creatively. It should eliminate bureaucratic obstacles and let people spend more time on real work—devising new ways to outmarket the competition,

serve the customer better, and allocate resources more intelligently.[32]

Few firms are farsighted enough, and perhaps successful enough, to make such investments. But InterAAct and other such investments in infrastructure processes do not themselves constitute process innovation; they may be necessary, but they are not sufficient. Management attention must be directed to converting the raw capabilities of infrastructure into the operational and management process innovations they enable. Otherwise, it becomes an expensive tool for doing the same work as before.

SUMMARY

We have argued here less for management process innovation than for management process. Management behavior in most organizations is so unstructured as to be difficult even to view as a process, much less as a target for process innovation. We have advocated, in specific process domains, gentle movement in the direction of process through the use of process description and measurement. These actions can be viewed as first steps toward a long-term goal of applying the approaches and techniques of process innovation to management behavior.

Until a process approach is adopted, we suspect that such innovations as executive information systems will continue to gather dust on rosewood credenzas. Activities, to be improved, must first be understood and regularized. Although technology and other enablers of innovation can benefit management activities, we cannot yet predict management behavior well enough to know how to help them do so.

Notes

[1]The source for these figures is *Employment and Earnings*, Bureau of Labor Statistics, U.S. Department of Labor, 1990.

[2]Most of Henry Mintzberg's research on managerial work is collected in *Mintzberg on Management* (New York: Free Press, 1989).

[3]This research was originally reported in Henry Mintzberg, *The Nature of Managerial Work* (New York: Harper & Row, 1973) and in Mintzberg, "The Manager's Job: Folklore and Fact," *Harvard Business Review* (July–August 1975): 49–61.

[4]John P. Kotter, *The General Managers* (New York: Free Press, 1982); Kotter's ideas and others are also usefully discussed in John F. Rockart and David

W. DeLong, *Executive Support Systems* (Homewood, Ill.: Dow Jones-Irwin, 1988).

[5]Daniel J. Isenberg, "How Senior Managers Think," *Harvard Business Review* (November–December 1984): 80–90.

[6]Masaaki Imai, *Kaizen* (New York: McGraw-Hill, 1986).

[7]For more information on policy deployment, see Yoji Akao, ed., *Hoshin Kanri: Policy Deployment for Successful TQM* (Cambridge, Mass.: Productivity Press, 1991).

[8]Kathleen M. Eisenhardt, "Speed and Strategic Choice: How Managers Accelerate Decision Making," *California Management Review* 32:3 (Spring 1990): 39–54.

[9]For examples of learning-oriented management styles, see Peter M. Senge, *The Fifth Discipline: The Art and Practice of the Learning Organization* (New York: Doubleday, 1990).

[10]On treating planning as a learning exercise, see Arie de Geus, "Planning as Learning," *Harvard Business Review* (March–April 1988): 70–74.

[11]For a recent and compelling discussion of the need for these measures, see Robert G. Eccles, "The Performance Measurement Manifesto," *Harvard Business Review* (January–February 1991): 131–137.

[12]David Harvey and Ian Meiklejohn, *The Executive Information Systems Report,* vol. 1 (London: Business Intelligence, 1990): 229–232.

[13]Rockart and DeLong, *Executive Support Systems,* 34.

[14]See references to computer-based information throughout Sharon M. McKinnon and William J. Bruns, Jr., *The Information Mosaic* (Boston: Harvard Business School Press, 1992).

[15]Richard Munton, ICI decision-support manager, quoted in Harvey and Meiklejohn, *The Executive Information Systems Report,* vol. 1, 249.

[16]Some aspects of Xerox's work in the corporate law department are described in Amy Dockser, "Xerox to Lose Its Legal Chief to Law Firm," *The Wall Street Journal,* November 17, 1988, B1.

[17]Senge, *The Fifth Discipline,* 181; de Geus, "Planning as Learning."

[18]John D. Sterman, "People Express: Management Flight Simulator," software and briefing book (1988). Available from the author, Sloan School of Management, 50 Memorial Drive, C52-562, Massachusetts Institute of Technology, Cambridge, MA 02142-1347.

[19]Personal conversations with Colgate-Palmolive executives, 1988.

[20]Apple Computer, "Business Navigator," 1989.

[21]Personal conversation with Dow Chemical information systems and financial executives, June and September 1991.

[22]Lynda M. Applegate and Charles Osborne, "Xerox Corporation: Executive Support Systems," 9-189-134. Boston: Harvard Business School, 1988, rev. 1990. This case describes the effort to simplify and standardize reporting structures, but not the return to the more diverse reporting environment.

[23]Eccles, "The Performance Measurement Manifesto"; also personal communication, July 8, 1991.

[24]Peter Krass, "Backlash at Big Blue," *InformationWeek* (July 29, 1991): 4–46.

[25]Harvey and Meiklejohn, *The Executive Information Systems Report,* vol. 1, 96.

[26]Lynda M. Applegate and Charles Osborn, "Phillips 66 Co.: Executive Information Systems," 9-189-006. Boston: Harvard Business School, 1988.

[27]Robert Wallace, Phillips Petroleum, cited in Rockart and DeLong, *Executive Support Systems,* 258.

[28]Applegate and Osborne, "Xerox Corporation: Executive Support Systems."

[29]See John Sviokla, "Texas Instruments: Capital Investment Expert System," 9-188-050. Boston: Harvard Business School, 1988.

[30]Lynda M. Applegate and Julie H. Hertenstein, "Westinghouse Electric Corporation: Automating the Capital Budgeting Process (A)," 9-189-119. Boston: Harvard Business School, 1989.

[31]The InterAAct system is described in detail in Joyce Wrenn, "InterAAct—Creation of a Technology Platform," in *Information Empowerment,* Critical Technology Report C-4-1 (Carrollton, Tex.: Chantico Publishing, 1991).

[32]Max Hopper, "Rattling SABRE—New Ways to Compete on Information," *Harvard Business Review* (May–June 1990): 118–125.

Chapter 14

Summary and Conclusions

In this concluding chapter, we look both backward and forward—backward to provide an overview of this broad and complex topic, and forward to implementation issues. Our view of implementation reflects a need for both caution and urgency. Caution is advised to avoid undertaking process innovation initiatives outside the organizational context or culture they require to be successful. On the other hand, there are numerous reasons to begin process innovation now. We explore this seeming conflict in implementation approaches in greater detail later in the chapter.

Overview of Key Messages

Six key messages for managers and analysts recur throughout this book. They constitute the key premises and conclusions of our work. Each has been "proven" to some degree by the experience of firms and their consultants. The messages needing further empirical proof are backed by high levels of logic and inference from current practice.

Our discussion takes these messages in no particular order, save that the first message is largely presumed by the others.

1. *Process innovation is a new and desirable approach to transforming organizations and improving their performance* (Chapter 1).

Process innovation, the radical improvement of business process performance through the use of innovative tools and work designs, has roots in the quality movement and other approaches to operational betterment of business activities. It combines these sources in a unique fashion.

Although the quality movement has developed the notion of processes and process improvement to a high degree, its orientation is to incremental rather than radical change, and it does not address enablers of change. We have argued for the combination of programs of process improvement and process innovation—applied to different processes at different times and for different reasons. We believe that a quality orientation that combines process improvement and process innovation efforts is unique and uniquely relevant to American and Western business. It is in keeping with our cultural leanings toward innovation and results, yet incorporates the rigor and measurement orientation found in the quality approaches of many successful Japanese firms. But because improvement and innovation are quite different, it is important to know which is pursued in a particular instance.

Process innovation, although admittedly difficult to achieve because of the radical nature of the organizational change involved, is a highly appealing approach to business transformation. It can be undertaken at relatively low cost, and the design, if not the implementation, of new processes can be completed in a matter of months. Efforts that yield tenfold improvements in selected aspects of process performance are not uncommon. For these reasons, many firms in the United States and Europe, in virtually all industries, are embarking upon major process innovation initiatives.

2. *An explicit approach to process innovation is important* (Chapters 2–8).

Process innovation does not take place in a casual, offhand manner. The pressures of day-to-day business and organizational inertia make a concerted and structured effort necessary if real change is to be achieved. The details of a specific method or approach to process innovation may vary, but the inclusion of several key activities is critical to the success of any initiative. These include selecting processes for redesign, giving structured consideration to enablers of innovation, creating a vision, understanding the existing process, and designing the new process and organization in detail.

3. *Information and information technology are powerful tools for enabling and implementing process innovation* (Chapters 3, 4, 10).

Although it is theoretically possible to bring about widespread process innovation without the use of computers or communications, we know of no such examples. IT is both an enabler and an implementer of process change. The relationship between IT and process-based structures is reciprocal; processes require information technology to achieve radical change, and to harness the capabilities of information technology in a cross-functional, performance-driven manner requires a process view.

Information technology can both provide opportunities for and impose constraints on process innovation. Opportunities take a variety of forms, including automational (eliminating human labor from a process), sequential (changing the sequence of a process or performing the tasks in parallel), geographical (enabling a process to operate effectively over great distances), and disintermediational (eliminating process intermediaries). IT can support virtually every step in the creation and implementation of new processes, as well as rapid and flexible creation of process support systems. IT-imposed constraints on innovation derive largely from aspects of existing IT infrastructure that cannot or will not be changed. Such constraints are found in every firm, and it is important to acknowledge them and determine how they limit the degrees of freedom in designing new processes.

Information is itself a powerful process resource. Although we are just beginning to understand how to manage information in a process context, already it is clear that accurate, real-time information on process performance is a prerequisite for effectiveness. Many processes have as their primary objective the creation of information.

4. *How a firm approaches organization and human resources is critical to the enablement and implementation of innovative processes* (Chapters 5 and 9).

How people are organized and managed and the degree to which they are empowered to do their work are critical to the success of process design. Specific organizational and human resource approaches likely to enable innovation in process design include empowering workers to handle entire processes (particularly at the customer interface), establishing autonomous work teams, and creating new, more process-oriented organizational structures. Such actions typically work in concert with information technology.

Organization and human resources are, in our experience,

frequently viewed as constraints to process innovation. Companies sometimes assess the receptiveness of their cultures to organizational change with the implicit assumption of cultural constraints.

Organization, like IT, is critical not only to enabling, but also to implementing, new process designs. In fact, if IT is the primary factor in enabling process innovation, organizational change is the primary factor in implementing it. Process innovation must be treated and managed as large-scale organizational change. Change roles must be clearly identified and change processes followed, and a strong awareness of change issues must inform all aspects of process innovation, from initial process visioning to prototyping and final implementation.

5. *Process innovation must occur within a strategic context and be guided by a vision of the future process state* (Chapter 6).

The formulation of a strategy and vision for process innovation must precede design and analysis. Lacking a strategic context, only incremental improvement achieved by eliminating bureaucratic or non-value-adding tasks is likely, and without a vision, it is hard to know what to innovate and what type of improvement to pursue.

A vision for process innovation should be closely tied to the organization's strategy. It should specify the processes to be redesigned, the attributes of those processes, and quantifiable process improvement objectives. Process designers and implementers must be provided specific targets toward which to direct their efforts. A tight connection between corporate strategy and process vision can render process innovation initiatives a primary vehicle for implementing strategy, and with strategy implementation becoming an important source of competitive differentiation, firms that are successful at process innovation are likely to be successful in the marketplace.

6. *Innovation initiatives can benefit all manner of processes* (Chapters 11, 12, 13).

We have devoted three chapters to process innovation for specific process types, ranging from product development and delivery (manufacturing and operations), to such customer-facing processes as sales, order management, and service, to various types of management processes. Our examples are taken from a variety of industries. Although we have concentrated on industries that began process innovation efforts early (i.e., information technology vendors, insurance, banking, pharmaceuticals, other manufacturing), we have found the potential for process innovation very high in many other industries as well. Only in industries in which workers tend to view their tasks as extremely unstructured and creative is process innovation particularly difficult. This is the case for professional services, investment banking, and scientific environments as well as for management processes generally. This is not to say that process innovation is not valuable or possible in these industries, only that it is more difficult to implement in the face of countervailing beliefs on the part of process practitioners.

We have also primarily addressed the private sector in this book. But there is no reason to believe that process innovation efforts coming to our attention in the public sector are any less likely to succeed. Of the many examples of successful use of IT in the public sector, some have a strong process orientation.[1] Some public-sector organizations, such as the U.S. Internal Revenue Service and State Department, and the Ontario Ministry of Revenue in Canada, have begun or completed major process innovation initiatives. Organizational change and acquisition of sophisticated information technology may be more difficult in some quarters of the public sector, but both occur every day.

REASONS FOR CAUTION IN PROCESS INNOVATION

Process innovation is a function more of management rhetoric than management science. This is not a pejorative observation; most management nostrums, however well buttressed by statistics and footnotes, fall into this category.[2] Rhetoric persuades through force of logic and oratory. In a new area such as process innovation, this is the only means of advancing the state of knowledge.

The hazard of rhetoric, however, is that some listeners will be overly persuaded or hear only the easy or appealing components of the message. A specific hazard for process innovation is that firms will undertake initiatives that do not fit their broader organizational context.[3] Process innovation must somehow be associated with a broad program of cultural change, whether it engenders that change or results from it. An early initiative can run counter to an organization's culture, but for it and subsequent initiatives to be successful, the culture must adapt.

The cultural climate in which process innovation takes place must be one in which operational business improvement is an important focus. The organization must be sufficiently disciplined to accede to measurement and analysis of results, and receptive to innovation, empowerment, and change. We have occasionally been told, "Quality initiatives didn't work in our organization, so we are planning to reengineer our processes." That probably will not work. Where there is little product innovation, there is little chance that process innovation will succeed. Where IT is not taken seriously as an enabler of competitive advantage, it will not be taken seriously as an enabler of process innovation.

Tom Peters has explained the logic behind moving toward greater levels of innovation and offered some vehicles for doing so.[4] David Nadler has summarized many of these cultural attributes as they relate to quality, but it is worth noting that they also relate to process innovation. We recount a few of what he labels "quality-hostile assumptions" below.

- We're smarter than our customers. We know what they really need.

- Our primary and overriding purpose is to make money, to produce near-term shareholder return.

- Our key audience is the financial market, in particular, the analysts.

- The primary way to influence corporate performance is through portfolio management and creative accounting.

- Managers are paid to make decisions. Workers are paid to do, not think.

- The job of senior management is strategy, not operations or implementation.

- If it isn't broken, don't fix it.[5]

Again, successful process change can be a factor in changing culture. There are many reasons besides a desire to accommodate process innovation for an organization to be open to innovation, change, a strong emphasis on IT, and empowerment. If an organization exhibits none of these characteristics, the process innovation enthusiast should go elsewhere or wait until there are significant signs that the organization is opening to change.

Companies, even those with cultures receptive to process innovation, should not expect to achieve major change without making major commitments. Successful process innovation relies on a wide range of skills. To effect needed change, organizations must somehow mobilize sufficient technological, human and organizational, political, and process expertise in concert with the requisite enablers. This necessarily involves assigning some of their best people, or, if the firm lacks the needed skills or methods internally, employing external consultants, to design and implement new processes. Absence of skills is as much a reason for caution as an unreceptive culture.

Throughout this book, we have emphasized that high levels of cross-functional coordination are needed to make processes work. High levels of cross-functional cooperation are essential in every stage of process innovation as well. Successful process design and implementation may rely on skills associated with information systems, human resources, finance, and general management, as well as with the process itself. Permanent groups of internal process innovation consultants, such as are being established by Kodak and CIGNA, among other firms, should draw members from information systems, quality, industrial engineering, and other backgrounds.

For any given process, the cycle of process design and implementation can easily take two years. Ford's increasingly well-known process innovation, which eliminated three-quarters of the accounts payable staff by paying on receipt of goods and bypassing invoices, took five years from concept to full cutover. IBM's goals for its 18 key processes are stated in multiyear increments, some goals reaching five years out.

Consequently, process innovation must itself be viewed as a process, not a project. If initial efforts are successful, companies will move on to redesign other processes, a prospect that stretches to decades. Absence of such long-term orientation is often decried in American business; process innovation is one more reason it cannot continue.[6] In financial terms, the best candidates for process innovation are those whose short-term survival is not in question, but whose long-term viability is unlikely without major change.

The long-term nature of process innovation leads to issues of executive continuity. Enthusiasm for process innovation must not reside with a single executive, however important that one person. For process innovation to succeed over the long term, a management group that transcends individual executives and organizational structures must be committed to process management, to information technology, and to organizational change. Process innovation is most likely to succeed in firms that exhibit a constancy of mission or "strategic intent" that is shared widely among executives and employees.

REASONS FOR HASTE IN PROCESS INNOVATION

Notwithstanding the risks attendant on implementing process innovation in less than the fully desired context, there are important reasons for throwing caution to the winds and launching a major initiative. Foremost is that process innovation is here to stay. There have been enough successes to indicate its feasibility and potential, and firms pursuing it may well be your competitors.

Process innovation is not a fad. Despite the rhetorical aspect of discussions about the topic, few would question the permanence of its foundations. The value of viewing a business cross-functionally, through the eyes of customers, is not a fad. The notion of combining radical and continuous improvement of business activity is beyond fashion. The idea that information technology and human factors can enable organizational change is perennial. Process innovation will be a concern of good managers for many years; what is remarkable is that it has taken so long to develop approaches to it.

A number of firms that began process innovation initiatives before there was any publicity about, or understanding of, the pitfalls and best approaches managed to be successful and benefited greatly from their efforts. Today, there are articles, books, consultants, research programs, training courses, and methodologies to assist the many companies undertaking process innovation initiatives.

Investment in the potential enablers of process innovation continues apace. Firms are spending millions of dollars to build systems that support product development and order management processes. These investments should support, not the tired old processes now in place, but innovative new process designs. Workers being empowered and trained to execute old processes should be learning about innovative new processes as well. As one senior IT executive in a large leisure-oriented firm put it, "I am no longer prepared to sign large checks to build systems and skills that will not improve our business performance; the only way I can avoid that is to insist on redesigning processes first."

So get going! The best way to learn about process innovation (after reading this book) is to try it. Begin to think of the organization as a collection of processes. Consider the many aspects of process innovation that must eventually be dealt with. Convince others of the value of the approach. Assemble a team, select a process, and begin to study it. Strive to create the best process of its kind in any business or industry. Spend no more money on building IT or human capabilities except in a process innovation context. Not to do so, as process innovation becomes the norm, is to forego not only an opportunity for competitive advantage, but even the ability to remain viable.

Notes

[1] For an overall discussion of IT-oriented innovation in the state of Minnesota, see Jerry Mechling and Judith A. Pinke, "Investing in Innovation: The Minnesota Approach," *MIS Quarterly*, forthcoming. For a single case study in the public sector, see Nitin Nohria and J. Chalykoff, "Internal Revenue Service: Automated Collection System," 9-490-042. Boston: Harvard Business School, 1990.

[2] For a discussion of the role of rhetoric in management, see Robert G. Eccles and Nitin Nohria with James D. Berkley, *Beyond the Hype: Rediscovering the Essence of Management* (Boston: Harvard Business School Press, 1992).

[3] This hazard applies to information technology and systems as well. See M. Lynne Markus and Daniel Robey, "The Organizational Validity of Management Information Systems," *Human Relations* 36:3 (1983): 203–226.

[4] Thomas Peters, "Get Innovative or Get Dead," *California Management Review* 33:2 and 3 (Winter and Spring, 1991): 9–26, 9–23.

[5] David A. Nadler of Delta Consulting Group, quoted in "Management Initiatives for Continuous Quality Improvement Programs," *I/S Analyzer* 29:2 (February 1991): 1–14.

[6] For a discussion of the short-term focus in U.S. business, and some suggestions on how it can be addressed, see Michael T. Jacobs, *Short-Term America: The Causes and Cures of Our Business Myopia* (Boston: Harvard Business School Press, 1991).

Appendix A

Companies Involved in the Research

Interviews or Discussions

North America
Ameritech
Apple Computer
AT&T
Bank of America
Bank of Boston
Baxter
Becton-Dickinson
Bethlehem Steel
Black & Decker
Bristol-Myers Squibb
Capital Holding
CIGNA
Citicorp
CoreStates Bank
Corning
Dow Chemical
Duke Power
Dun & Bradstreet
Eaton's
Faxon
Federal Express
Fidelity Management
   & Research
First Chicago
General Dynamics
General Electric
General Motors
Hallmark
Hewlett-Packard
Huntington Bank
IBM Canada
Intel
Kodak
Merck
McDonald's
Morgan Stanley
NCR
Nynex
Pacific Bell
Panhandle Eastern
Pioneer Standard

PNC Financial
Procter & Gamble
R.R. Donnelley
RJR Nabisco
Rochester Telephone
Scott Paper
Sherwin Williams
Silicon Graphics
Simon & Schuster
Society Bank
So. N.E. Telephone
Standard & Poor's
Syntex
Toys 'R' Us
United Parcel Service
Unum
USAA
Westinghouse

Europe
Air Products Europe
Bass PLC
British Petroleum
British Telecom
Honda (Europe)
ICL

Latin America
IBM Latin America
Pequiven (Venezuela)
Polar (Venezuela)

Japan
Japan Development
   Bank
Mitsui
Toshiba

Extended Analysis

North America
Continental Bank
Digital Equipment
Federal Mogul
Bethlehem Steel
Mutual Benefit Life
IBM
Westinghouse
Xerox (Corporate, U.S.
   Marketing Group)

Europe
Rank Xerox U.K.

The Origins of Process Innovation

Process innovation, though a new concern of organizations, has roots that can be traced well back into the middle of this century. The impulse to improve operational performance has been a lasting concern of the twentieth century, manifested in today's process innovation methods and techniques. The notions of adopting a process orientation and of making order-of-magnitude leaps in performance have flourished in the past in the West and elsewhere. Even the notion of using IT and human enablers of change to benefit operational activities is not innovative.

What is new is the combination of these elements in a well-defined approach to process innovation. The concerns are timeless and the components well tested. Process innovation is only an extension of ideas that have been employed frequently over the past several decades. This is evidence, we believe, that process innovation is not a fad. How to structure and measure business activities and occasionally achieve radical performance improvements are enduring concerns of business. Organizations, now that they have the tools and methods to bring about this type of change, will be dealing with process innovation for many years to come.

Process innovation has its origins in a variety of approaches to business improvement. Primary sources include the quality movement, industrial engineering and systems thinking, the work design approaches pioneered by the sociotechnical school, analysis of the diffusion of technological innovation, and ideas about the competitive use of information technology. Each domain will be discussed briefly, beginning with the quality movement. Although these business improvement notions are Western in origin, and all but the sociotechnical school are primarily American, they have been adopted in different forms by organizations in different geographies. Whenever possible, we explore these differences between Eastern and Western firms' adoption of ideas related to process innovation.

PROCESS IMPROVEMENT AND THE QUALITY MOVEMENT

Process thinking originated with the quality movement. Its focus on outputs and customers is consistent with the emphasis of early quality thinkers on minimizing variation and defects in manufactured products.[1] These experts argued strongly that processes should be stabilized and variation closely measured through statistical process control. After stabilization, steady but incremental improvement in the process could begin.

Radical process innovation was encouraged by some quality experts,[2] but not even mentioned by many others.[3] Edwards Deming was among those who held a mixed view—admitting the possibility of radical improvement, but arguing that a corporate culture that did not support continuous incremental improvement would not support more radical change.[4] Both Deming and Joseph Juran addressed innovation early in their careers as quality experts, but, perhaps because they felt American managers already leaned toward product innovation, their later work tended to emphasize disciplined, continuous process improvement.

Many Japanese firms have embraced cultures of continuous process improvement. Japanese managers tend to be extremely conscious of their process management responsibilities and can sometimes even cite the percentage of time they devote to process management.[5] Lester Thurow argues that Japanese firms spend two-thirds of their research and development budgets on new processes and only one-third on new products—the reverse of American firms' investment proportions.[6]

Indeed, some Japanese firms, and the quality experts whose precepts they follow, have emphasized process over result.[7] To the quality faithful, if the process is managed, the result takes care of itself. This philosophy seems to have worked well in Japan, and we are reluctant to raise the specter of national character differences. Yet, in the American business culture—noted for an emphasis on large, exciting change[8] and visible results—a process orientation that shuns results thinking may meet substantial resistance. According to a Japanese quality expert, Masaaki Imai:

> Japanese companies generally favor the gradualist approach and Western companies the great-leap approach—an approach epitomized by the term innovation. Western management worships at the altar of innovation. This innovation is seen as major

changes in the wake of technological breakthroughs, or the introduction of the latest management concepts or production techniques. Innovation is dramatic, a real attention-getter. *Kaizen*
[continuous improvement], on the other hand, is often undra-
matic and subtle, and its results are seldom immediately visible.
While *kaizen* is a continuous process, innovation is generally a
one-shot phenomenon.[9]

Imai and other leading thinkers about quality in both Japan
and the West often seem to denigrate innovation, to view it as
a "quick-fix" approach lacking in discipline and long-term benefit. The implicit assumption is that innovation is all right as
long as one does not strive consciously for it; home runs should
be hit by batters trying for singles. On the other hand, many
Western critics of Japanese firms have argued that they succeed
by capitalizing on the product and process innovations of Western
firms—e.g., the computer, the VCR, continuous casting of steel,
and even statistical process control. A recent survey suggests
that Japanese firms are highly interested in developing new (and
presumably innovative) processes and in using information technology to do so.[10] Clearly, businesses in all parts of the world
need both discipline and innovation, both process and result. The
ideal baseball team would include both Pete Rose, who reliably
delivered singles, and Henry Aaron, who could frequently deliver
a home run.

Some experts have begun to question the quality movement's
exclusive emphasis on process, and argue that American companies must forcefully demand, in addition to better processes,
radically better results from their managers and workers. One
observer noted that continuous improvement and total quality
programs "serve frequently as convenient escape mechanisms for
managers avoiding the struggle of radically upgrading their organizations' performance."[11]

Many American companies we studied have found it difficult
to instill a culture of continuous process improvement. Even companies making major commitments to the total quality management concept, including Ford,[12] Xerox, and General Motors,[13]
have encountered strong cultural resistance. At Florida Power
and Light, the only American company to win the prestigious
Deming Award, the CEO who spearheaded the firm's rigorous
efforts to become process-focused was recently relieved of his responsibilities.[14] Interviews with the firm's employees revealed an

overemphasis on the process and bureaucracy, rather than the results, of quality. The new CEO, in a memo, put it thus:

> These employees believe that undue emphasis on process deprives them of time that could be better spent serving customers and participating in community affairs . . . I was most troubled, however, by the frequently stated opinion that preoccupation with process had resulted in our losing sight of one of the major tenets of quality improvement, namely, respect for employees. Many employees mentioned that there is too little personal responsibility for results, and that there is less recognition for making good business decisions than for following the QI [quality improvement] process.[15]

Finally, many companies that embark on quality programs find it difficult to measure the financial rewards that derive from such programs.[16] Continuous, incremental improvement is often hard to quantify—but the costs of training, consulting fees, and other aspects of quality programs are more easily assessed. To remedy this imbalance, quality gurus have encouraged managers to assess the "costs of poor quality," though accepting the reality of these costs is sometimes a matter of faith. Moreover, the assessed cost of poor quality measures only costs associated with defects in existing products and processes, not the opportunity cost of failing to adopt process innovations.

PROCESS INNOVATION AND SCIENTIFIC MANAGEMENT

The industrial engineering and systems movement constitute another source of process innovation concepts. The earliest precursor of these ideas, the scientific management movement of the late nineteenth and early twentieth centuries, was based on the assumption that behavior at work could be engineered, designed according to principles of rationality and efficiency. The same approaches that had been successfully applied to technology were transferred to workers.

The primary apostle of this revolution, of course, was Frederick Winslow Taylor. Taylorism, as scientific management came to be called, included several key principles that are related to process innovation concepts:

- the separation of the execution of work from its design;

- the assumption that there is an ideal design for any work process;

- the need for measures and controls of work efficiency and effectiveness; and

- the need for workers to follow standard, routinized procedures.[17]

These Taylorist threads were used to weave other tapestries of business improvement. The function of industrial engineering was based to a large degree on Taylorist concepts. Industrial engineers were trained to break down work activities into small steps, measure them at intervals to the second, and recombine them in the most efficient manner possible.

When the defense industry began to build complex missiles during World War II, Taylorist principles were further refined into "systems thinking." Work was viewed as a complex system with measurable inputs and outputs, which required rigorous controls and measures. Process flow diagramming was developed to describe such work systems. As their complexity grew, demanding mathematical techniques for queue analysis, optimal routing, and so forth, the discipline of operations research was born.

When computers entered the business environment in the early 1950s, they were first applied to operations research problems. Many industrial engineers and operations researchers became systems analysts, a new role intended to spearhead the application of computers to business problems, that is, to design both work and computer systems. Given the enormous potential of even the earliest computers, the emphasis for many of these work designers shifted from squeezing seconds out of a work step to integrating the computer into work. As described by Robert Howard, the computer was a boon to the Taylorist.

> Seventy years later, the principles of Taylorism are alive and well in the assumptions of many technology managers. Indeed, one might even say that they are giving Taylor's idea of a dominant "system" new substance by building it into the hardware and software of computer-based work systems. Whereas Taylor wanted to subordinate man to this system, his successors today see computer technology, first and foremost, as a means to eliminate, or at least minimize, the human element in work. Just as Taylor called for the thorough standardization and routinization of tasks, they see computerization as a mechanism to

spread that systematic standardization into areas of the workplace and industries where it has rarely been seen before.[18]

Despite these intellectual roots, process innovation as we have
defined it is by no means a simple extension of Taylorism. The
human contribution to work is to be celebrated and optimized
rather than eliminated. Innovative process designs can leave room
for creativity and worker autonomy. Work can be designed as a
soft, human system rather than a rigid mechanical one.[19] The act
of process design can be a participative activity, although there
are often limits to the level of participation that is possible. Finally, the role of technology in process innovation is not simply
to minimize and standardize human labor, but to serve as a cat-
alyst for processes that would never have been possible without
it. Process innovation offers the possibility of making whole again
jobs that have become fragmented and specialized—of creating
jobs in which a worker, with the aid of IT, monitors and acts
across an entire organization. These pages contain numerous examples of work processes that illustrate the possibility of going
well beyond Taylorism.

PROCESS INNOVATION AND WORK DESIGN

The study and improvement of work processes have been the
focus not only of engineers, but also of practitioners of organizational behavior and development. The usual goal of the latter
was not improved process efficiency or the elimination of human
labor, but the creation of meaningful jobs and greater worker
satisfaction (the implicit or explicit assumption, of course, being
that worker satisfaction leads to higher productivity). Organizational theorists also focused on individual jobs and workers in
the context of a broader organizational and social environment.[20]

Beginning in the 1940s, several different "schools" of organizational behavior contributed to thinking about processes and
work design. One, led by Kurt Lewin, focused on improving human interactions within small work groups.[21] Another work group-
oriented school, led by Elton Mayo and Fritz Roethlisberger of
the Harvard Business School, tried to understand the implications
of social interaction for small group performance.[22] Ironically,
their ideas were refined in the telephone industry (at Western
Electric), where statistical process control was being developed
at roughly the same time.

The role of technology in work design began to be addressed by a group in England and the United States called the sociotechnical school. Sociotechnical work design is the harmonious blending of technology and people in a structured work process, taking into account the culture and environment in which a work process activity occurs. Although it is difficult to object to the principles of sociotechnical work design, the literature contains few detailed guidelines for combining human and technological resources in a satisfactory way for specific work environments.

The sociotechnical approach has been applied in many industrial environments, frequently as a way to enrich the jobs of assembly-line workers or to introduce technology effectively into unionized job environments.[23] But this approach has stalled over the last decade, because the assumption that more satisfied workers would become more productive workers has not always been realized. What process innovation can bring to sociotechnical work design is the importance of radical improvement in results and the measurement and improvement focus of process thinking. What sociotechnical design brings to process innovation is just as important: the need for concurrent management of technological and human change, with a strong focus on human enablers of innovation.

PROCESS INNOVATION AND THE DIFFUSION OF INNOVATION

Some aspects of the sociotechnical mantle have been assumed by researchers investigating the diffusion and implementation of innovation in organizations, sometimes called "technology transfer." The best examples of this research, seen in the work of Rogers,[24] Gibson,[25] and Leonard-Barton,[26] view innovation in the broader context of organizational behavior and change. Many of the innovations studied involve information technology, and some involve innovations in process.

Within this tradition of management research, analysis of the diffusion of innovation usually begins with the design, and continues through implementation, of the innovation. The goal of this research is to understand, not how innovations can improve organizational performance, but how they can be successfully adopted. Successful implementations of innovation result from the mutual adaptation over time of both technology and organizational structures and processes.

This perspective is useful, particularly in managing the introduction of new information technologies. Some of the findings of researchers who have analyzed the interplay between technological and human resource enablers of innovation are useful for understanding the nature and difficulty of process innovation.[27] Nord and Tucker, for example, found that banks attempting to implement a new type of checking account encountered the most difficult environments when the innovation had to span functional boundaries.[28]

But the primary focus of research on organizational innovation has been on technologies, both product- and process-oriented. Business processes as innovations have not been an object of study. We can learn from the difficulty of implementing other types of innovations, however. "The implementation of organizational innovations, of which new technologies are a subset," observes Leonard-Barton, "is an extremely complex management problem, requiring much skill."[29]

PROCESS INNOVATION AND COMPETITIVE IT

With quality, systems, work design, and diffusion of innovation concepts well established by the 1980s, only one other conceptual advance was necessary for process innovation to become possible—the notion that information technology could be a source of dramatic change and business improvement, that is, that information technology could provide competitive advantage.

In the late 1970s and early 1980s, a number of companies began to employ computers and communications not just to automate back-office processes, but in frontline roles with customers. Airlines' reservation systems penetrated travel agencies; hospital supply and drug distribution firms placed terminals on customers' premises; suppliers of heating and air conditioning equipment provided customers with software to help them establish equipment needs as they designed their buildings. These systems, and the ideas behind them, are described in a series of articles from the early and mid-1980s,[30] but there were many other, less advertised, successes.

Neither these articles nor the managers who attempted to implement competitive advantage through IT emphasized that these stories of competitive success involved considerable change in business processes. The successful firms may not have had an explicit process orientation, but they clearly were taking advantage of IT to enable process innovation in marketing, order management, and integrated logistics processes. The systems they implemented were cross-functional, had an impact on customers, and yielded measurable improvements in operational performance. However tiresome these widely circulated strategic systems legends might have become, they acquainted managers in all walks of business with the idea that IT can radically alter both the way work is done and the resulting balance of competition.

The supply of competitive IT legends dried up quickly, however. Few firms feel today that they have gotten competitive advantage from their IT investments. Of many possible reasons for this, one obvious candidate is the absence of a process orientation in many firms. Included as an element of such a firm's strategy, IT might not measurably improve implementation of the strategy. Had they embraced the discipline and results orientation that characterize a process view of the organization, many more firms might have capitalized more fully on their investments in strategic IT.

The five conceptual streams described above are important tributaries to the river that is process innovation. Just as these approaches provided ideas that made the notion of process innovation possible, so process innovation provides a means to address problems associated with each of these approaches. Quality, systems thinking, work design, diffusion of innovation, and competitive IT have provided process innovation with methods, enablers, and objectives. Another key aspect of process innovation, a radical pace and degree of change, also has a history, as we shall see.

CHANGES IN THE PACE OF CHANGE

Thus far in this Appendix we have discussed different intellectual precursors to process innovation. The degree of change assumed or sought by these different approaches has been addressed only tangentially. Explicit treatment of the level of change will clarify the distinction between process innovation and previous approaches to operational improvement, particularly those based on quality. The radical approach to business improvement proposed in this book, the most radical of three major approaches to the level and pace of business change that have evolved over the past 90 years, is but a logical extension of product and process quality efforts begun early in this century.

The first major instance of formal business improvement was the use of product inspection as the final step in a manufacturing process (see Figure B-1). This activity, again formalized by Taylor as an essential task in manufacturing,[31] involved little change in the process itself; poor-quality goods were simply rejected or recycled, with little investigation as to cause or prevention. In fact, no process view of the organization was taken; industry had only recently adopted the functional, bureaucratic division of labor that enabled mass production.[32]

Beginning with the pioneering efforts of Shewart, Deming, and other quality researchers at Bell Laboratories in the 1930s, a second approach to business improvement was developed. This approach, frequently called "quality control," involved strict analysis and control of the production process for manufactured goods. Although the focus was on the manufacturing process (or, in the telephone industry, on service operations) rather than cross-functional processes, at least production activities were viewed from beginning to end. Variation in the process was to be measured and minimized through statistical analysis, that is, statistical process control.

Figure B-1   The History of Process Improvement Approaches

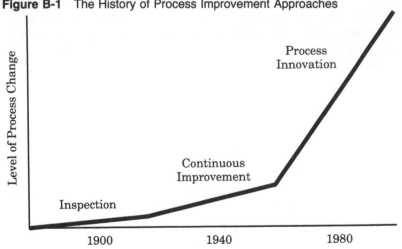

The quality control notion has been greatly expanded from the early days at Bell Labs, although its derivative concepts still dominate the quality field. The application of quality control has been expanded from manufacturing processes to all processes involved in producing quality products and services: product design, supplier relationships, and logistics.[33] Deming[34] and Crosby,[35] together with other experts, expanded the concept of quality control to include implications for management style, human resource policy, and other wide-ranging issues.

Yet even the focus of total quality management (the term now attached to the broad view of quality) remains on making continuous improvement and minimizing variation in existing processes.[36] The desired level of change is incremental. The focus on statistical measures and product-oriented (as opposed to service and administrative) processes continues to this day.

The possibility of radical process innovation, as we have noted, was admitted, sometimes grudgingly, by a few quality experts, but only when it occurred within a continuous improvement context. Furthermore, no leading thinker about quality has seized upon the process improvement opportunities offered by IT, except as a means of collecting and manipulating process variation statistics.

By the mid- to late 1980s, many companies in the United States and Europe had begun to suspect that continuous improvement was insufficient to meet their needs for business change. Companies such as Xerox, IBM, Ford, CIGNA, Bank of America, Kodak, Hallmark, and Bethlehem Steel, among many others, recognizing the need for more radical change in their business processes, initiated programs of process innovation with the objective of achieving major reductions in process cost and execution time, or realizing major improvements in the quality of process outcomes.

Just as they were first to adopt notions of continuous improvement, manufacturing companies, according to Hayes and Wheelwright, have been among the first to pursue breakthrough-level business improvements.[37] These companies refer to the differences between continuous improvement and innovation as "tortoise versus hare" approaches to competition. They suggest, as Nadler does,[38] that innovation-oriented approaches are better suited to the national character of U.S. managers, incremental improvement to that of Japanese and German managers. The concept of process innovation combines the methodical, deliberate nature of the tortoise with the speed and ambition of the hare.

Even in companies that have made radical improvements in their manufacturing processes, there is often considerable room for improvement and innovation in staff functions and administrative processes. More than 40% of jobs in the manufacturing sector involve office work, which the quality movement has left relatively unchanged,[39] particularly in service companies.

In the insurance industry, for example, many of the work structures for processing insurance policy applications were broken up into smaller and smaller jobs. In the interest of increasing volume and simplifying work, the processes became enormously complex. The designers of these processes never considered, and did not measure, the communications and buffer time consumed by a process that involves many functions and many jobs. Such processes took a long time, required many people, employed outdated systems, and did not serve their functions very well.

In the steel industry, for example, although great improvements were made in casting technology and basic manufacturing processes, the inefficiencies described above persisted in the processes that linked customer demands with production scheduling. Functional boundaries and information systems that fail to cross those boundaries make it very difficult for most steel companies to commit to producing a given customer order at a given time for a given price. The process usually involves several iterations between customer, sales representative, and production scheduling department, with the result that the cycle time to fill an order is generally measured in weeks and months, rather than days.

In some cases, initiatives to address these poorly structured processes were driven by competitive and other business demands. When customers, for example, demanded more synchronized deliveries, the steel industry, unable to afford large inventories, saw little alternative to process innovation.

In other cases, the primary impetus for process innovation comes from an information systems function that believes the technologies it supports to be capable of enabling radical improvement. This is not a new phenomenon; the IS function has long held out its systems as tools for changing business, but until the late 1980s, the function had lacked the requisite respect and credibility. IS personnel found that satisfying customer needs increasingly required building systems to support processes rather than business functions or other types of structural units. Even today, however, the IS function may not be the ideal advocate of process innovation. Many managers remain suspicious of technology, and IS does not own the key processes that need to be changed.

Toward the end of the 1980s, the idea of redesigning or reengineering business processes gained popularity. Two articles on the topic of process innovation and the role of IT,[40] one co-authored by the author of this book, elicited a strong positive reaction from IT and business executives. A number of surveys of IT executives in the early 1990s confirmed that process innovation with IT was a high priority. Today, many consulting firms, both technology and general management-oriented, offer process innovation as a service.

Although process innovation programs are quite different from continuous improvement programs in the same companies, both arise out of the same tradition. They share an orientation to process as the unit of improvement, an orientation to strategy execution rather than strategy itself, a belief in the importance of measurement and analysis, and a focus on external criteria (e.g., benchmarking) as the basis for judging improvement. When continuous and radical improvement initiatives coexist within an organization, they should be perceived by the organization's members as being related aspects of an overall operational improvement initiative.

SUMMARY

All of these precursors of process innovation share one important concern—they are oriented toward operational improvement, that is, they assume that business advantage comes not through better strategies, but through doing business differently. Otherwise, they differ in many respects; they propose different levels of change, focus on different aspects of the innovation process, and emphasize different enablers of change. Each makes a distinct contribution to process innovation thinking.

The quality movement brings to process innovation an emphasis on processes and process betterment. Although it does not have a strong emphasis on innovation (as we noted, some quality experts are hostile to radical change), the quality movement's continuous improvement approach complements process innovation.

The principles of scientific management espoused by Taylor, and later woven into the disciplines of industrial engineering and systems analysis, are firmly linked to process innovation, having shown a strong emphasis on the decomposition of work, as well as a focus on measurement and, eventually, on the power of technology to enable change in work. But they neglected radical change and the role of human and organizational factors in work motivation and design.

The sociotechnical school was the first to combine human and technological enablers of change. In this sense, process innovation as we have defined it is a sociotechnical approach. But sociotechnical analysis has traditionally lacked a strong operational and financial improvement focus, and there is no well-defined sociotechnical method. Consequently, its primary use has been to explain or minimize problems in existing work environments, rather than to help design new ones.

The diffusion of innovation literature, particularly some of the more recent work, brings a valuable perspective to process innovation. Its focus on the implementation of innovation and the mutual adaptation of organization and technology is highly relevant to how process innovation can be successfully implemented. To date, its primary focus has been on product innovation, or relatively narrow innovations in manufacturing processes. As more firms implement process innovation, they will become a fruitful area of study for diffusion of innovation theorists.

Finally, the competitive use of IT brought several important dimensions to process innovation. This school of management research (and rhetoric) emphasized IT's ability to dramatically change the way companies do business. One problem with competitive systems thinking was the absence of a process context, or any other structured approach, for achieving operational improvement from IT initiatives. In general, the focus was on identifying IT-enabled strategies, rather than on using IT to implement strategies. With IT-enabled process innovation, companies are clearly achieving strategic advantage.

Notes

[1]David A. Garvin, *Managing Quality* (New York: Free Press, 1988): Chapter 1.

[2]Joseph M. Juran, *Managerial Breakthrough* (New York: McGraw-Hill, 1964).

[3]See, for example, Kaoru Ishikawa, *What Is Total Quality Control? The Japanese Way* (Englewood Cliffs, N. J.: Prentice-Hall, 1985).

[4]Quoted in Andrea Gabor, *The Man Who Discovered Quality: How W. Edwards Deming Brought the Quality Revolution to America—The Stories of Ford, Xerox & GM* (New York: Random House, 1990).

[5]Personal conversation with Larry Ford, former head of IBM Asia.

[6]Lester C. Thurow, "New Tools, New Rules: Playing to Win in the New Economic Game," *PRISM* (Cambridge, Mass.: Arthur D. Little, Second Quarter 1991): 87–100.

[7]See, for example, Masaaki Imai, *Kaizen* (New York: McGraw-Hill, 1986): 16–21.

[8]James Brian Quinn, *Strategies for Change: Logical Incrementalism* (New York: Richard D. Irwin, 1980).

[9]Imai, *Kaizen*, 23.

[10]See "Factories of the Future," Executive Summary of the 1990 International Manufacturing Futures Survey, Boston University (1991). The survey found that Japanese companies rate developing new processes as a higher priority than American or European firms.

[11]Robert H. Shaffer, "Demand Better Results—and Get Them," *Harvard Business Review* (March–April 1991): 142–149.

[12]For a discussion of the early years of Ford's quality initiatives, see David Halberstam, *The Reckoning* (New York: William Morrow, 1986); for an internal view, see Donald E. Petersen and John Hillkirk, *A Better Idea* (Boston: Houghton Mifflin, 1991).

[13]On Xerox and GM, see Gabor, *The Man Who Discovered Quality*.

[14]Robert Chapman Wood, "A Hero Without a Company," *Forbes* (March 18, 1991): 112–114.

[15]Memorandum from James Broadhead, quoted in Jerry G. Bowles and Joshua Hammond, *Beyond Quality* (New York: Putnam, 1991): 149.

[16]There are exceptions to this, however. Motorola claims that it saved $942 million in three years through improved quality. See Garrett DeYoung, "Does Quality Pay?," *CFO* (September 1990): 24–34.

[17]Frederick W. Taylor, *Principles of Scientific Management* (New York: Harper & Row, 1911).

[18]Robert Howard, *Brave New Workplace* (New York: Viking, 1985): 19.

[19]For a detailed discussion of soft systems approaches, see Peter Checkland and Jim Scholes, *Soft Systems Methodology in Action* (Chichester, England: John Wiley, 1990).

[20]Richard Hackman and Greg Oldham, *Work Redesign* (Reading, Mass.: Addison-Wesley, 1980).

[21]See, for example, Kurt Lewin in Dorian Cartwright, ed., *Field Theory in Social Science: Selected Theoretical Papers* (New York: Harper & Row, 1951).

[22]See, for example, Elton Mayo, *The Human Problems of an Industrial Civilization* (New York: Macmillan, 1933); see also F. J. Roethlisberger and William J. Dickson, *Management and the Worker* (Cambridge, Mass.: Harvard University Press, 1939).

[23]For an example of sociotechnical thinking at Volvo, see P. G. Gyllenhammar, *People at Work* (Reading, Mass.: Addison-Wesley, 1977).

[24]Everett M. Rogers, *Diffusion of Innovations*, 3d ed. (New York: Free Press, 1983).

[25]Cyrus F. Gibson, "A Methodology for Implementation Research," in R. L. Schultz and D. P. Slevin, eds., *Implementing Operations Research/Management Science* (New York: American Elsevier, 1975): 53–73.

[26]See, for example, Dorothy Leonard-Barton, "Implementation as Mutual Adaptation of Technology and Organization," *Research Policy* 17 (1988): 251–267.

[27]See Urs E. Gattiker, *Technology Management in Organizations* (Newbury Park, Calif.: Sage Publications, 1990).

[28]Walter R. Nord and Sharon Tucker, *Implementing Routine and Radical Innovations* (Lexington, Mass.: Lexington Books, 1987).

[29]Dorothy Leonard-Barton, "Implementation Characteristics of Organizational Innovations," *Communication Research* 15:5 (October 1988): 603–631.

[30]See Gregory Parsons, "Information Technology: A New Competitive Weapon," *Sloan Management Review* (Fall 1983): 3–14; F. Warren McFarlan, "Information Technology Changes the Way You Compete," *Harvard Business Review* (May–June 1984): 98–103; Michael E. Porter and Victor E. Millar, "How Information Gives You Competitive Advantage," *Harvard Business Review* (July–August 1985): 149–160.

[31]Garvin, *Managing Quality,* 5.

[32]Max Weber, *Theory of Social and Economic Organization* (New York: Free Press, 1947).

[33]Feigenbaum as cited in Garvin, *Managing Quality,* 254, ff. 28.

[34]Gabor, *The Man Who Discovered Quality.*

[35]Phillip B. Crosby, *Quality Is Free: The Art of Making Quality Free* (New York: McGraw-Hill, 1989).

[36]See, for example, H. J. Harrington, *The Improvement Process: How America's Leading Companies Improve Quality* (New York: McGraw-Hill, 1987).

[37]Robert H. Hayes and Steven C. Wheelwright, *Restoring Our Competitive Edge* (New York: John Wiley, 1984).

[38]David Nadler, "America, Playing to Its Own Strengths," *New York Times* (June 2, 1991) Financial Desk, late edition, sec. 3, p. 11, col. 2.

[39]Raymond R. Panko, "Is Office Productivity Stagnant?," *MIS Quarterly* (June 1991): 191–203.

[40]Thomas H. Davenport and James E. Short, "The New Industrial Engineering: Information Technology and Business Process Redesign," *Sloan Management Review* (Summer 1990): 11–27; also Michael Hammer, "Reengineering Work: Don't Automate, Obliterate," *Harvard Business Review* (July–August 1990): 104–112.

Index

Aaron, Henry, 313
Accounting information for management, 77, 143
ACD, *see* Automated call distributor
Action diagrams, 147
Activity value analysis, *see* Value analysis
Aetna, 276
Aetna Life Insurance, 97, 108
Aguillar, Frank, 79
AI, *see* Artificial intelligence
Akers, John, 290
Allaire, Paul, 5
American Airlines, 54, 122, 207, 295
  InterAAct platform, 295, 296
  SABRE reservation system of, 254
American Express, 53, 88
  Authorizor's Assistant of, 53
  IDS division of, 269
American Hospital Supply, 39
American President Lines, 62, 249
American Telephone & Telegraph (AT&T), 43, 88, 111, 126, 253
Ameritech, 15
  major processes of, 7
Ancona, Deborah Gladstein, 99, 184
ANI, *see* Automatic number identification
Artificial intelligence (AI), 170
Athos, Anthony, 118
AT&T, *see* American Telephone & Telegraph
Automated call distributor (ACD), 51
Automated design, 56–57, 60
  as generic application, 56–57

Automatic call distributors, 103–104
Automatic number identification (ANI), 255

Balcor, 103
Baldrige award, 10, 117, 190
Barclay's Bank, 269
Bass plc, 294
Baxter/American Hospital Supply, 253–254
Baxter Healthcare, 122, 249
Becton Dickinson, 76
Beer, Michael, 162, 190
Bellcore, 63–64
Bell Laboratories, 320–321
Benchmarking, 55, 63, 125, 126
Bethlehem Steel, 34
Black & Decker, 153
Bowen, William, 45
Brainstorming, 150, 154, 156
Bristol-Myers Squibb, 294, 295
British Petroleum, 276
British Telecom, 28, 43, 276
Bulletin board
  electronic, 58, 103, 269
Businesses, integrator, 260
Business management
  global, 18

CAD, *see* Computer-aided design
Caldwell, David E., 99
Camp, Robert C., 126
CANDA, *see* Computer-Aided New Drug Application
Capital Holding, 276
Carlzon, Jan, 262
CASE, *see* Computer-aided software engineering
Case management, 102, 104, 107, 130, 260–265

327

About the Author

Thomas H. Davenport is a partner in Ernst & Young's Center for Information Technology and Strategy in Boston, Massachusetts. He is responsible for the research and multiclient program activities of the center and also participates in selected consulting engagements. Prior to joining Ernst & Young, he was director of IT research at McKinsey and Index Group. He has taught information technology management at the Harvard Business School, Boston University School of Management, and the University of Chicago. Davenport has published widely in the IT and general management press, including the *Harvard Business Review, Sloan Management Review,* and *Management Review.* He is also the co-author of *The Information Imperative.*